IIE Research Report Number Twenty-one

D1499252

choosing futures:

U.S. AND FOREIGN STUDENT VIEWS OF GRADUATE ENGINEERING EDUCATION

ELINOR G. BARBER, Columbia University
ROBERT P. MORGAN, Washington University
WILLIAM P. DARBY, Washington University

Acknowledgments

The authors gratefully acknowledge the assistance of many individuals in carrying out the study. Foremost among these are Laura J. Sallmen-Smith and Robert M. Sholtes. We also thank the Alfred P. Sloan Foundation for providing the financial support that made the study possible. Certain sections of this article were published previously in E.G. Barber, R.P. Morgan, W.P. Darby, and L.J. Sallmen-Smith, "To Pursue or Not to Pursue a Graduate Engineering Degree," *Engineering Education* , Vol. 79, No. 5, July/August 1989, pp. 550-555.

Contents

Introduction

In the modern world, it is of great importance for any country to have an adequate supply of highly-trained scientists and engineers; in a country like the United States, an adequate supply of well-trained graduates of engineering schools is widely believed to be a very significant element, not only in national security and economic competitiveness but also in ensuring a satisfactory standard of living and quality of life of the country's citizenry. In developing countries the issue is often not the need for engineers but how and where they may most appropriately be trained.

For the United States, it is very important to have a sufficient supply of graduate engineers, and there is considerable support for the position that the current and projected total supply falls short of meeting demands. It is quite clear that there is and will very likely continue to be a shortage of U.S. citizens with advanced engineering degrees, particularly at the doctoral level. These advanced degrees are essential for effective research in industrial laboratories and for teaching and research in engineering schools. Therefore, it is of some consequence that there now appear to be too few U.S. citizens who take the advanced degrees required to staff most positions in academia and certain positions in industry. The slack is being taken up by foreign nationals who obtain these advanced degrees from U.S. engineering schools. One telling statistic is that in the mid-1980s, half of all assistant professors in U.S. engineering programs younger than 35 years of age were not U.S. citizens.[1] (See References, page 55.)

In the 1980s, the annual percentage of foreign graduate students in U.S. engineering programs had exceeded 40% and the proportion of doctorates awarded to non-U.S. citizens had consistently exceeded 50%. From the point of view of U.S. engineering schools, this influx of foreign graduate students is certainly a plus. As Barber and Morgan have shown elsewhere,[2,3] the flow of foreign students is perceived by chairpersons and faculty of U.S. engineering programs as an asset: with fewer foreign students, it would be impossible to sustain research and instruction at existing levels. Many foreign graduate students are of high quality and perform essential roles as research and teaching assistants. Their English language competency causes some communication difficulties and their strong orientation to theory may not be congenial to all faculty, but still they are deemed to be indispensable assets.

The resulting composition of the graduate student population does, however, cause some uneasiness in many quarters, largely because of the extent to which engineering schools and the engineering labor market have become dependent on the continuing influx of foreign students. A large majority of these students come from India, Taiwan, and the People's Republic of China. It should suffice to allude to the recent upheaval in the PRC and its potential for disruption of the continuing inflow of Chinese students to appreciate the validity of concerns about such dependence.

There are various possible remedies. One is to reduce the number of engineering graduate programs, perhaps eliminating the weaker ones. This option is not being prominently discussed, partly because the very idea of eliminating programs causes great uneasiness, and partly because such a drastic measure might well exacerbate the projected shortage of engineers. The second remedy, less frightening but so far elusive, is to significantly increase the number of U.S. citizens in graduate engineering programs.

If the proportion of U.S. graduate students, particularly doctoral candidates, is to be increased, it is first necessary to understand the reasons why U.S. engineering undergraduates decide to be or not to be graduate students. Further, it is necessary to understand the reasons why those, U.S. or foreign, who do go to graduate school in engineering decide to become candidates for doctoral degrees. An adequate flow of U.S. citizens through the pipeline into faculties of engineering schools or into industrial laboratories will come about only if appropriate numbers of students in particular disciplines, female as well as male, minority as well as white, make the necessary academic choices. If these choices were better understood, appropriate interventions might be designed to increase the likelihood of positive choices. Put another way, if science and technology policymakers knew the reasons why foreign students frequently make positive choices and U.S. students do not, appropriate interventions might be fine tuned. And these interventions might be adjusted even more delicately if more were known about the motivations of underrepresented groups, i.e. women and minorities within the U.S. population of engineering students.

The most common explanation of the decisions by U.S. students with baccalaureate degrees in engineering *not* to pursue further training is couched in economic terms. The explanation goes like this: a first degree in engineering is generally sufficient for the recipient to get a well-paying job; the income foregone or deferred by graduate study is not recouped by the higher earnings that an advanced degree permits. These economic arguments are surely cogent; there is indeed evidence of their cogency in the higher enrollments in graduate training that occur at times of declining job opportunities. To the extent that these economic arguments are important, there may be economic measures that could increase graduate enrollments: larger fellowships and higher salaries for those obtaining advanced degrees. The question remains whether there are also *other* important factors influencing decisions to be or not to be a graduate student or a candidate for a Ph.D. A significant percentage of U.S undergraduate engineering students do continue their studies in engineering and many foreign students with B.S. degrees pursue advanced engineering degrees.

If economic reasons alone were so compelling, why would anyone go to graduate school? What other factors might be involved? We hypothesized a number of factors which we assume affect different kinds of undergraduates in different ways. These include:
- the educational experiences of some undergraduate engineering students and especially the possible "burn-out" phenomenon attendant on very demanding undergraduate programs;
- the process of recruitment into graduate engineering programs, and a lack of encouragement by faculty and of adequate information about academic offerings and financial aid;
- the negative perceptions some undergraduates have of graduate study and especially of doctoral study in engineering;
- the career aspirations of certain undergraduate engineering students and the role that graduate study plays in their fulfillment;
- the perception, especially but not only by women, of engineering as a profession for white males;
- the sense of some undergraduates that the engineering profession lacks glamour and visibility and that the work is not challenging or socially fulfilling;
- the lack of support for graduate study that some undergraduate engineering students receive from their families;
- the attractiveness to some graduate students of employer-financed graduate study to the master's, but not the doctoral, level.

Methodology

To find out more about the pertinent decisions of engineering students we conducted two nationwide surveys in the spring of 1988, one of full-time seniors, the other of full-time graduate students* in U.S. engineering programs.[4] We obtained agreement from 479 chairpersons of chemical, civil, electrical, and mechanical engineering departments to distribute the questionnaires either in a classroom setting or individually. The sample was stratified according to three variables: discipline; governance of the institution; and the quality and/or research intensiveness of the department. With regard to this last variable, two strata, namely QRI-1 and QRI-2, contain departments that were rated in the top and bottom halves, respectively, of a 1982 assessment of U.S. research-doctorate programs in engineering.[5] A third stratum, QRI-3, contains departments that were not included in the 1982 assessment but were listed in the most recent American Society for Engineering Education (ASEE) compilation of engineering research and graduate study[6] or were added as undergraduate programs accredited by the Accreditation Board for Engineering and Technology (ABET). Thus, QRI-3 departments tend to include either new departments, those that produce relatively few doctoral degrees, or those that offer no graduate programs; they tend to be less research-intensive than QRI-1 or QRI-2 departments, but their relative quality is harder to gauge.

For our survey we randomly selected a number of departments within each stratum and asked chairpersons of the departments to survey all of the students in them. This was deemed the most efficient and least expensive method of conducting the study.

We oversampled several kinds of departments: those in the top-20 engineering schools, as rated by engineering deans in 1987; those that were the top-20 Ph.D. producers in 1986; the top-20 producers of bachelors degrees in engineering in 1986; and the top-20 producers of bachelors de-

---

* We did not survey part-time students; in engineering, they are not of considerable importance, particularly at the master's level.

grees awarded to women, blacks, and Hispanics, respectively, in 1986. However, the oversampling did not significantly skew the results based on usable responses received.

We also oversampled Ph.D. candidates against M.S. candidates at the graduate level. We took this approach because of the strong interest in the dearth of U.S. citizens going on to the doctoral level. However, this means that in considering the total responses of graduate students the reader should keep in mind that doctoral candidates are overrepresented.

We aimed at achieving survey responses from 4,500 students (2,000 seniors, 1,500 masters, and 1,000 doctoral students). A total of 22,836 questionnaires were mailed out and a total of 4,880 usable questionnaires were received (2,416 from seniors and 2,464 from graduate students). Of the 773 chemical, civil, electrical, and mechanical engineering departments and

Table 1

Comparison of Distribution of Survey Respondents
with Overall Degrees Awarded, by Gender, Citizenship,
and Race/Ethnicity

Respondent	Surveyed % Undergraduate	B.S. Degrees Awarded (1987)*%	
Female	17.7	15.4	
Foreign	9.4	8.0	
U.S. blacks	1.7	2.9	
U.S. Hispanics	3.3	3.0	
U.S. Amer. Indian	0.9	0.16	
U.S. Asian Amer.	5.7	6.6	
Respondent	**Surveyed % Graduate**	**Graduate Degrees Awarded %**	
Female Students	13.4	12.8	M.S.
		7.1	Ph.D.
Foreign Students**	42.6	26.0	M.S.
		43.1	Ph.D.
U.S. blacks	0.7	1.5	M.S.
		0.4	Ph.D.
U.S. Hispanics	2.1	1.6	M.S.
		1.0	Ph.D.
U.S. Amer. Indian	0.3	0.14	M.S.
		0.14	Ph.D.
U.S. Asian Amer.	7.5	7.3	M.S.
		5.6	Ph.D.

* Source for degree data, *Engineering Education*, May, 1988.
** *Engineering Education* data tend to underestimate degrees awarded to foreign citizens.

programs in the United States, questionnaires were sent to 226 and survey data were obtained from students in 174, for a departmental response rate of 77%, and the large total mailing of student questionnaires produced an estimated student response rate of at least 21% from the seniors and 22% from the graduate students.

The 21%–22% estimated response rate represents a lower bound of the actual student response rate. Our procedure involved sending department chairpersons as many questionnaires as we estimated the sample population to be, or more in the case of small sample departments. Some of the mailed questionnaires may never have been distributed. If we knew how many of these there were, we could calculate the effective response rate more accurately; in any event, it would be higher than 21%–22%. Given our relatively modest resources for carrying out this project, we feel reasonably satisfied with our approach to obtaining responses.

For the majority of the subpopulations of interest, the distribution of responses in the survey was reasonably close to the distribution in the actual student population, indicating that the majority of our results are representative (see Table 1). We did find in analyzing our responses that there was little deviation from representativeness on the governance or quality variables, but some deviation on the disciplinary variable. We therefore weighted the responses to correct for this deviation and compared the weighted and unweighted results. We found that the unweighted results were sufficiently similar to the weighted ones to permit us to base our conclusions on unweighted results.

Findings: Engineering Seniors

It is important to keep in mind some demographic characteristics of the engineering seniors who responded to the questionnaire. First of all, 90.6% of the seniors are U.S. citizens, a vastly higher proportion than in the case of graduate students. Of the seniors 17.2% are female and 11.5% come from minority groups. About half of this latter category are Asian, 1.7% are black, 3.3% are Hispanic, and 0.9% are Native American. Clearly, even though we oversampled students from the top-20 schools awarding bachelors degrees to blacks and Hispanics, our sample contains a small number of respondents from these groups. Only the blacks, however, are significantly underrepresented among minority respondents (see Table 1).

The questionnaire administered to seniors had three major components dealing with: 1) their undergraduate education, 2) their post baccalaureate plans and the reasons for those plans, and 3) their career aspirations and views of the engineering profession, as well as a section asking for personal and family background data. In a later section we shall compare the characteristics of those U.S. seniors headed immediately for Ph.D.s and those planning never to go to graduate school, as well as the characteristics of some of the other subgroups mentioned above. In the next sections, we shall focus on the variable of academic ability in analyzing the responses of engineering seniors, contrasting the responses of students with GPAs in the highest quartile with those in the lowest quartile.

Undergraduate Education

The judgments seniors make about their undergraduate engineering programs presumably have significant bearing on their disposition to seek further education in this field. From this perspective, it would be especially important to find out if sizable proportions of students found their undergraduate education to have been, for example, too demanding or theoretical or very boring. Further, it is especially interesting to compare the overall pattern of senior responses to questions about their undergraduate education with the responses of specific subgroups: the more and the less academically-able students and the students planning to go to graduate school. There is much concern that the best and the brightest are not being attracted to graduate study, in the first instance, and to doctoral study and academic careers, in the second; our data suggest that these fears may be unwarranted.

Of the responding seniors, 38% said that they found their undergraduate engineering program excessively demanding (they feel "burned out") (see Table 2), and this percentage rises to 48% among those seniors who have no intention of ever going to graduate school. The academically-abler students with the GPAs in the highest quartile, however, were less likely to complain of burn-out — 31% of them felt their program had been too demanding, compared to 46% of the students with GPAs in the lowest quartile. Similarly, while 31% of the seniors said they found their undergraduate engineering program too theoretical, among those with the highest GPAs the percentage was only 25%. Yet even a burn-out rate of 31% among high-GPA students is not low: it would be very useful to explore further why these able students found their undergraduate programs too demanding and what factor this judgment played in their decisions to continue or not to continue their studies.

Table 2
Assessment of Current Undergraduate Program

Frequency	%	Frequency	%	Frequency	%
Too Theoretical 734	30.8	About Right/Neutral 1,600	67.2	Too Practical 46	1.9
Too Specialized 119	5.0	About Right/Neutral 1,858	78.5	Too Broad 390	16.5
Very Stimulating 424	17.9	About Right/Neutral 1,627	68.6	Very Boring 316	13.3
Too Demanding: I Feel Burned Out 900	37.9	About Right/Neutral 1,396	58.8	Not Demanding Enough 72	3.0
Overall, Very Satisfactory 717	30.1	About Right/Neutral 1,437	60.4	Overall, Not Very Satisfactory 222	9.3
Reinforcing My Desire To Be An Engineer 878	37.1	About Right/Neutral 1,211	51.1	Convincing Me That I Don't Want To Be An Engineer 277	11.7

Percentages shown are based on response to a three-point item in a horizontal row.

Post-Baccalaureate Plans

Only 19% of all seniors graduating in the spring of 1988 in our sample planned to go to graduate school in the fall of 1988, which means that the overwhelming majority had no immediate plans to do so; 60% of those continuing planned to obtain a master's degree and 33% a doctorate.

Table 3
Importance of Reasons for Attending Graduate
School in Engineering

| Reason | Very Important | | Somewhat Important | | Rank |
	Frequency	%	Frequency	%	Very/Somewhat
The subject matter is interesting and challenging	237	70.5	90	26.8	2 1
I do well in academic work	129	38.5	147	43.9	5 3
I need a graduate degree to do the work that interests me	166	49.4	109	32.4	3 4
I want to pursue a career in research	98	29.3	105	31.3	7 6
The life of an engineering professor seems attractive to me	48	14.3	84	25.0	9 9
I have been offered or expect a good financial aid package	109	32.5	77	23.0	6 7
I want to improve my future salary prospects	162	48.4	107	31.9	4 5
I have no attractive full-time job offers	23	6.8	49	14.6	12 12
I am not ready to go to work	47	14.0	84	25.0	10 10
I want to learn more in depth about my field	242	72.0	72	21.4	1 2
In general, my family regards advanced degrees highly	72	21.4	100	29.8	8 8
Many of my friends have, or intend to get, advanced degrees	26	7.8	62	18.5	11 11
Other (Please specify)	16	19.3	5	6.0	13 13

Furthermore, of the 19% planning to go to graduate school, only four out of five (81%) planned to stay in engineering. Although 73% of those who were not immediately going to graduate school thought they might eventually do so, the most important question this skewed distribution raises is how to increase the dismayingly small proportion that now view graduate school and doctoral degrees positively.

It is with regard to the reasons for opting for or against graduate education that the relative importance of economic and non-economic factors is of central interest. If we examine only very important reasons for pursuing a graduate degree (see Table 3), the most frequently offered ones could be called idealistic: the challenging, interesting subject matter and a desire to learn more in depth about a field. Contrary to conventional wisdom, economic considerations do not predominate, but they are far from unimportant. Almost half the seniors thought graduate education would improve their future salary prospects. It should be kept in mind that the economic advantages of alternative choices may not be fully understood by undergraduates, though among those *not* planning to go to graduate school immediately, a sizable percentage of responses (29%) indicated that their most important reason was the offer of a well-paying job in industry.

It is interesting to note what might induce this latter group to go eventually to graduate school for engineering: 75% indicated that they might do so if their employer paid for their education and 65% indicated that the encouragement they might receive from employers would be a significant factor. The economic factor surfaces in another way: according to 23% it would make a difference if jobs in their engineering field became more scarce or paid less. For 24% graduate education would become more attractive if academic jobs became better paid (see Table 4). For a good many seniors

Table 4
Possible Reasons for Attending Graduate School Later

Reason	Frequency	%
I need or want to learn about my engineering field	1,188	83.0
My employer pays for my graduate education	1,076	75.2
My employer encourages me to get some graduate training	935	65.3
Academic jobs become better paid	337	23.5
Jobs in my engineering field become more scarce or pay less	329	23.0
Fellowship or assistantship stipends are increased	199	13.9
Research facilities in universities are improved	131	9.2
My undergraduate loans are forgiven	123	8.7
Other (Please specify)	86	6.0

Percent based on total responding to at least one item on the list.

who have had it with school, the condition can be reversed by various economic incentives.

Among those students who thought they might eventually go to graduate school, those with the highest GPAs emerge as more likely to consider doing so: 81% of those in the highest quartile answered positively, compared to 64% in the lowest quartile.

Regarding perceptions of doctoral study, the positive views tend to be couched in intellectual terms, the negative ones in economic or occupational ones (see Table 5). According to the positive views, the degree provides a strong theoretical basis for an engineering career (59%) and doctoral studies provide access to state-of-the-art equipment and facilities (44%); also, it would be an intellectually rewarding experience (34%).

Table 5
Expectation of Doctoral Studies in Engineering

Expectation	Frequency	%
Provide access to state-of-the-art equipment and facilities	1,009	44.3
Provide a strong theoretical basis for my engineering career	1,341	58.8
Be much more difficult than undergraduate engineering	1,116	48.9
Be too much work for too small a future payoff	575	25.2
Overqualify me for industry	558	24.5
Require that I specialize too much	784	34.4
Be stimulating; I'd learn much that was interesting	778	34.1
Be necessary for success in a technical field	179	7.9
Be useful for only highly research-oriented people	1,067	46.8
Other	58	2.5

One response is ambiguous, depending on the academic strength of those making it: for more able students, the fact that doctoral studies are much more difficult than undergraduate engineering might be a plus, for weaker students, a minus; in any case this response was given by 49% of respondents.

The unambiguously negative responses are that doctoral studies would require too much specialization (34%); and that they would entail too much work for too small a pay off (25%) and overqualify the degree recipient for industry (25%). It is possible that effective career counseling could persuade able students that their negative arguments should be reconsidered.

About two-thirds of the seniors were discouraged by the length of time they thought it took to get such a degree. This may be in part an economic concern, while only about one-third were discouraged by the prospect of producing a dissertation. Since the seniors' estimates of the time it takes to get a doctorate were, if anything, on the low side (42% estimated four years of full-time study after the bachelors degree and 22% estimated five years; see Table 6), it would not be possible to influence their decisions by pointing out that their estimates are exaggerated. Concern in policy circles about the long period of time it takes to get a doctorate in all fields is growing.

Table 6
Estimate of Time to Doctoral Degree

Number of Years	Frequency	%
One year	10	0.4
Two years	104	4.4
Three years	391	16.7
Four years	990	42.3
Five years	517	22.1
More than five years	203	8.7
Don't know	126	5.4
Total	**2,341**	**100.0**

The students who are headed for graduate school tend to be among the ablest students: 32% of those in the respondent population who planned to continue their education fell into the highest GPA quartile, compared to 9% in the lowest GPA quartile. Our data show that, not surprisingly, it is these good students who are receiving strong encouragement from faculty at their undergraduate school: 63% of them reported encouragement, as did 51% of the students in the second-highest quartile, but only 28% of the students in the lowest quartile received such encouragement (see Table 7). Nor is it surprising that among those with high GPAs, 63% offered as a reason for going to graduate school that they do well in academic work, while only 4% of the lowest GPA students offered that reason. Able students are, obviously, both self-selected and encouraged to go to graduate school. While there are surely very able students among the 81% in the respondent population who decide against advanced education, any efforts to encourage more participation in graduate education need to be made with full awareness that a good many of those who do not continue may know their own limitations. This sense of academic and occupational identity comes across rather clearly in our discussion below of the differences between those headed for Ph.D.s and those who want never to go to graduate school.

Table 7
Sources of Encouragement for Graduate Study

	Overall Encouraged Me %				Overall Discouraged Me %				Neutral or Not Applicable %			
GPA Quartile	1	2	3	4*	1	2	3	4	1	2	3	4
Engineering faculty at undergrad. school	28.4	37.8	51.6	63.3	8.5	5.5	3.7	2.5	63.1	56.7	44.7	34.2
Non-engineering faculty at undergrad. school	7.6	10.5	11.2	12.7	2.6	3.0	1.9	3.4	89.8	86.5	86.9	83.9
Graduate teaching assistants at undergrad. school	16.9	19.1	23.4	26.5	6.0	6.5	4.8	5.5	77.0	74.4	71.8	68.1
My family	35.2	36.3	36.8	42.0	6.4	10.6	9.5	8.7	58.4	53.0	53.7	49.4
My fellow students	21.9	22.1	24.4	29.3	12.3	12.9	13.9	13.4	65.8	65.1	61.7	57.4
My friends	24.0	24.0	23.5	30.7	11.8	11.7	11.1	10.4	64.2	64.3	65.4	58.9
My spouse	7.3	8.6	7.3	9.6	4.2	6.0	5.5	4.8	88.5	85.4	87.3	85.6
My employer	16.2	21.1	18.7	21.4	3.1	3.3	5.5	5.5	80.8	75.6	75.8	73.1
Other	4.7	3.2	3.1	6.3	3.6	5.9	4.1	3.7	91.7	90.9	92.9	90.0

* 4 is highest GPA quartile, 1 is lowest.

The most frequently offered reasons for *not* attending engineering graduate school were being tired of going to school and a wish to earn a living or to work at least for a while before attending graduate school (see Table 8). Being tired of going to school is not, in itself, an economic reason; it may reflect the exhaustion or discouragement of the less able student. But it may also be a frame of mind in which economic reasons assume greater cogency. Twenty-seven percent of the seniors who turn away from graduate education said that they had incurred large debts for their undergraduate education; as noted above, 29% said they had been offered well-paying industrial jobs; and 13% had received no financial aid or insufficient financial aid for graduate education. It is important to note that 31% felt that more education is attractive but not economically justifiable.

Table 8

Reasons for Not Attending Graduate School

Reason	Frequency	%*
I have had enough of school	1,089	56.4
The kinds of jobs that require master's degrees or doctorates don't appeal to me	231	12.0
I have been offered a job in industry that pays well	555	28.8
I have been offered a job with part-time graduate opportunities	152	7.9
More education will not lead to a significantly better job	272	14.1
The life of a professor does not seem attractive to me	388	20.1
More education is attractive but not economically justifiable	597	30.9
I have incurred large debts for my undergraduate education	527	27.3
I have not been offered any financial aid for graduate school or the amount offered is too small	252	13.1
Most of my friends who can get good jobs are taking them	119	6.2
Engineering is a profession for white males	20	1.0
Graduate engineering education is dominated by foreign students and faculty	175	9.1
I was dissatisfied with my undergraduate engineering education	255	13.2
My family thinks it's time I earned a living	248	12.8
I think it's time I earned a living	1,024	53.1
In general, engineering professors have discouraged me from going to graduate school in engineering	63	3.3
I have not been accepted by any engineering graduate schools	86	4.5
I plan to work for a while, and then do graduate engineering work	830	43.0
Graduate programs, especially doctorates, take too long and outcomes are too uncertain	177	9.2
I don't think I can handle graduate work in engineering	160	8.3
I found a field I like better than engineering	141	7.3
Other	324	16.8

* Percent of responses; multiple responses allowed.

Since it would be very useful to know what it would take to make these seniors change their decisions, we asked what minimum annual fellowship stipend (in addition to fully paid tuition, and after taxes) might induce them to undertake full-time engineering graduate studies. We found that 23% of the seniors thought $10,000 would suffice, 19% stipulated $15,000, and 12% said $20,000; only 7% said financial aid would not be an important factor in their decision and 16% said they did not know (see Table 9).

Table 9
Minimum Fellowship/Stipend to Induce Interest
in Graduate School

Annual Stipend Amount	Frequency	%
Less than $5,000	59	4.2
$5,000	126	8.9
$10,000	331	23.3
$15,000	271	19.1
$20,000	172	12.1
$25,000	62	4.4
More than $25,000	70	4.9
Financial aid wouldn't be an important factor	93	6.6
Don't know	233	16.4
Total	**1,417**	**99.9***

* Differs from 100% due to round off error.

Images of Engineering and Career Aspirations

Whether or not engineering seniors go to graduate school may be related to their images of the kinds of rewards engineering offers — intellectual achievement, a comfortable living, self-fulfillment, or a sense of service. No very clear patterns emerge in this regard. A very high proportion of engineering seniors have a positive view of the engineering profession as useful (87%), challenging (75%), and important for the future of the United States (69%; see Table 10). None of those positive characteristics, except perhaps the first, is closely linked with the necessity for graduate study. A high proportion of the seniors also thought engineering pays well (68%) and leads to success in business (40%), and only 8% thought it was too closely tied to corporate interests. These latter images of the engineering profession have a strong economic component, but graduate education may not be required to realize them. Strongly negative images are not prominent: 10% saw engineering as too oriented towards military work and 10% saw it as too dominated by white males. These images of the profession may not be such as to dispose students to go to graduate school. However, it might be possible to recruit more graduate students by demonstrating that many of the positive qualities of engineering are enhanced if engineers are equipped to deal with more difficult or sophisticated problems.

Overall, the engineering activities the seniors most frequently hope to participate in are design and development, followed by management and administration, and then by consulting (see Table 11). Research is cited by 29% and teaching is a preferred activity for only 21% of the respondents, and teaching at the university level by only 12.4%. We do not know to what

Table 10

Impressions of Engineering Profession

Impression	Frequency	%
Exciting and challenging	1,763	74.9
Boring	201	8.5
Engineers do many useful, practical things to benefit mankind	1,920	81.6
Too oriented towards military work	242	10.3
Too dominated by white males	234	9.9
Important for the future of the United States	1,621	68.9
Pays well	1,596	67.8
Too closely tied to corporate interests	188	8.0
Does not project a clear, visible image	236	10.0
Serves society and the public interest	1,190	50.6
Leads to success in business	950	40.4
Other	92	3.9

Table 11

Preferred Engineering Activities After Graduation

Activity	Frequency	%
Design and development	1,667	70.9
Management and administration	1,196	50.9
Consulting	994	42.3
Production and/or manufacturing	853	36.3
Research	685	29.1
Communicating with the public	527	22.4
Teaching	502	21.4
Sales and/or marketing	462	19.7
Other (Please specify)	57	2.4
Don't know	40	1.7
None	26	1.1

extent the seniors realize what role graduate education plays in preparation for these various activities, though it seems likely that they are reasonably well-informed.

The seniors in the highest GPA quartile were more than twice as likely to be interested than those in the lowest quartile in research (42% vs. 20%) and in teaching (31% vs. 14%), while those in the lowest GPA quartile were almost twice as likely to be interested in sales and/or marketing (25% vs. 13%). The high-quartile GPA students also had a somewhat stronger preference for design and development than the low-GPA students (76% vs. 65%). Since these high-GPA seniors are also more likely to go to graduate school, it seems probable that the abler students have a better grasp of the academic requirements for different types of activities.

The career aspirations of the seniors are of some consequence in considering the possibilities of attracting more of them into graduate education (see Table 12). The top aspiration is to make a comfortable living (77% of responses), and the next ones are to make significant practical and technical contributions to a field and to manage important projects. Presumably, if it could be demonstrated to the seniors that advanced training is likely to enable them to combine these aspirations, i.e., to work on significant, practical projects and to make a *more* comfortable living than they would with only a bachelors degree, a higher proportion of them might opt for graduate education.

Table 12
Career Aspirations: All Undergraduate Respondents

Aspiration	Frequency	%
To make a comfortable living	1,789	76.5*
To make significant, practical, useful technical contributions in my field	1,527	65.3
To manage important projects	1,332	56.9
To contribute to the well-being of my fellow citizens	1,037	44.3
To start my own company	856	36.6
To help improve living conditions for those less fortunate both here and abroad	694	29.7
To work for a large company	671	28.7
To make significant research contributions to furthering knowledge	453	23.2
To make a product and sell it	476	20.3
To work for a small company	446	19.1
To teach at the university level	289	12.4
To work in federal, state or local government	185	7.9
To teach at the secondary school level	148	6.3
To work for a public interest group	81	3.5
To join the Peace Corps	73	3.1
Other (Please Specify)	57	2.4

* Percent of responses; multiple responses permitted.

Foreign vs. U.S. Undergraduates

Although the proportion of foreign undergraduates among engineering senior respondents is small** (9%), it is useful to know whether their responses shed any light on the patterns of participation of U.S. and foreign students at the graduate level.

** We defined foreign students as non-U.S. citizens studying on either temporary or permanent visas.

In at least a few respects, noticeable differences do show up between foreign and U.S. undergraduates. Most important, the percentage of foreign seniors who plan to go to graduate school is twice as high as the percentage of U.S. seniors (36% vs. 17%).* A higher proportion of foreign seniors majored or minored in math or science in addition to their engineering major (14% vs. 9% for U.S. seniors); a strong math and science background appears to be characteristic of students who go on for doctoral degrees. The foreign students also tend to have more friends majoring in engineering, science, and math than the U.S. students, suggesting that the foreign students have focused more narrowly on both technical and scientific interests and acquaintances during their undergraduate years.

Beyond that, foreign students indicated that they had received encouragement to go to graduate school from a far broader array of significant individuals: 65% of foreign students were encouraged by their family, as compared to 35% of U.S. students; by their friends (48% vs. 24% of U.S. students); by fellow students (35% vs. 23%); by graduate teaching assistants at their undergraduate school (32% vs. 21%); and by non-engineering faculty at their undergraduate school (22% vs. 10%). It is possible that foreign students are, or appear to be, more insecure and therefore solicit and elicit encouragement. It is also possible that U.S. undergraduates actually need a good deal more encouragement than they are receiving but do not feel comfortable about soliciting such support.

With respect to the reasons for attending graduate school, two major and quite different ones emerge between foreign and U.S. seniors. While only 10% of U.S. seniors said they had no attractive full-time job offers, as many as 33% of foreign seniors offered this reason; conversely, 30% of U.S. seniors vs. 13% of foreign seniors offered as a reason for *not* going to graduate school that they had been offered a well-paying job in industry. Either by choice or necessity, U.S. seniors seem more influenced by economic considerations in deciding not to continue their education. The economic pull away from graduate school appears to be stronger for the U.S. students and the pull towards graduate school appears to be quite a bit stronger for foreign students. Along with this pull, foreign students also receive more of a push: 40% of foreign students reported that their families highly regard advanced degrees compared to only 27% of U.S. students. It should also be noted that for foreign students, going to graduate school is a way of prolonging their U.S. sojourn; U.S. students have no such incentive.

U.S. students appear to have more reason not only to stay away from graduate school but also to defer advanced training. Some 78% of U.S. seniors

---

* To be sure, these foreign seniors constitute only 9% of all undergraduates surveyed, i.e., a small absolute number.

thought they might eventually continue their education if their employer pays for it, while only 42% of foreign seniors thought this might be an option for them.

Finally, in terms of career aspirations, U.S. engineering undergraduates are more oriented towards making a comfortable living than foreign undergraduates (77% vs. 54%); and — this is a bit inconsistent — the U.S. students are also more disposed to contribute to the well-being of their fellow citizens (46% vs. 30% for foreign students; see Table 13). An important finding from the point of view of this study is that foreign seniors more frequently indicated an interest in university-level teaching than did U.S. seniors (20% vs. 12%), though this activity seems not to be very high on the career agenda of either group; and, the foreign seniors were somewhat more interested in making significant research contributions (30% vs. 23% for U.S. seniors).

Table 13
Career Aspirations: U.S. vs. Foreign Undergraduates

Aspiration	% U.S.	% Non-U.S.
To make significant, practical, useful technical contributions in my field	65.7	61.1
To make significant research contributions to furthering knowledge	22.7	29.6
To manage important projects	57.6	50.3
To make a product and sell it	20.3	20.7
To make a comfortable living	78.6	54.2
To contribute to the well-being of my fellow citizens	45.7	30.0
To help improve living conditions for those less fortunate both here and abroad	29.1	34.5
To work for a large company	28.4	31.0
To work for a small company	19.7	12.3
To start my own company	35.7	44.3
To work in federal, state or local government	18.0	6.4
To work for a public interest group	3.2	2.5
To join the Peace Corps	3.2	2.5
To teach at the university level	11.6	20.2
To teach at the secondary school level	6.1	7.9
Other (Please specify)	2.6	1.5

The self-reported mean GPAs were fairly similar for the two groups, with the foreign students having slightly higher overall GPAs (3.13 vs. 3.06 for U.S. students) and GPAs in engineering courses (3.14 vs. 3.09), and lower for courses outside engineering (3.15 vs. 3.20), all based on a modified 4.0 scale. This would lead one to expect similar proportions of the two groups to continue their education, insofar as GPAs are a significant factor. But we did not directly compare foreign and U.S. seniors with the same GPAs with regard to their specific reasons for attending graduate school or their specific career aspirations. The overall patterns do not provide insight into the different ways in which foreign and U.S. students with comparable academic ability perceive the relationships between advanced education and the achievement of post-educational aspirations.

U.S. Racial/Ethnic Groups: Underrepresented (Black, Hispanic, American Indian), Asian-American, and White

In spite of our oversampling of institutions with the largest enrollments of blacks and Hispanics, the number of responses received from students in these racial/ethnic groups and from American Indians was very small. Therefore, we combined the 122 responses by U.S. students from these minority groups, whose proportion in the engineering student population is much smaller than their proportion in the general population, and called this the "underrepresented" group. The 118 Asian-American students were treated as another separate group. We make no claims for the representativeness of our findings, but because of the strong concern to recruit more members of underrepresented minority groups into undergraduate and graduate engineering programs, our findings may, at the very least, point the way to further exploration of the relevant attitudes and decisions of these underrepresented students.

A high proportion of underrepresented students who responded to the survey were in civil engineering (37%), compared to 27% of white students and 16% of Asian students; and only 15% of underrepresented students majored in electrical engineering, compared to 19% of white students and 48% of Asian students. Although this distribution may not be representative, these choices of majors by underrepresented students are consonant with their lower GPAs (overall 2.88), as well as with the higher GPAs of the Asian students (3.10) and white students (3.06): in the undergraduate engineering population, students with better academic records, as measured by GPA, are found in electrical engineering (3.22 average), and those with weaker records in civil engineering (2.96 average).

In spite of these differences in academic performance, it is rather surprising and interesting to find that roughly the same percentage of all three groups were planning to go to graduate school in the fall of 1988. Even more surprising and interesting, in all three groups a higher percentage than average of underrepresented students were expecting to work towards doctoral degrees. This may reflect either strong motivation on the part of underrepresented students or unrealistic expectations.

Before entering engineering school, a larger percentage of the underrepresented group (27%) went to private or parochial high schools than did Asian students (17%) or white students (15%). As undergraduates, underrepresented students were more involved in student government, community service, tutoring, and student chapters of professional societies than were white students. It is a matter of speculation whether their deep involvement in non-academic activities might have proved detrimental to their academic performance.

In addition to lower GPAs, other factors militate against the participation of underrepresented students in graduate education. The underrepresented students reported receiving less encouragement to continue their education than did the other two groups from engineering faculty members at their undergraduate school, teaching assistants, and their families, and more encouragement from fellow students, spouses, and employers. At the same time, these students also seem to have more financial need and to have to overcome more financial obstacles to graduate study. With regard to reasons for *not* going to graduate school, underrepresented students were less likely to respond that they had been offered well-paying industrial jobs than white or Asian students and more likely to express concern about their ability to handle graduate work. Significantly more of those in the underrepresented group (29%) than of Asians (7%) and of whites (14%) said they might go to graduate school at a later time if fellowship stipends were increased. A higher percentage of underrepresented students cited having incurred large debts for their undergraduate education as a reason for not going to graduate school (underrepresented 35%, Asian 24%, white 29%).

Among those planning to go to graduate school but not planning to obtain a doctorate, the underrepresented students were less deterred by the length of time a doctorate required (44% vs. 62% for whites and 71% for Asians) and more deterred by the dissertation requirement (67% for underrepresented vs. 30% for whites and 29% for Asians). We must point out, however, that these findings are based upon a very small number of responses (9 for underrepresented concerning the dissertation).

It is not easy to account for the patterns of choice of underrepresented students. Further research is needed to find out what it means to underrepresented students to go into engineering in the first place. We do know that fewer underrepresented students have engineering relatives (36%) than white students (53%) and Asian students (66%). We do not know whether an undergraduate engineering education has the same significance for them that it has for white or Asian students, and more specifically, whether and how their racial background was a factor in their choice of engineering. Insights of this kind would seem to be a prerequisite to understanding choices at the next higher educational level.

There are inconsistencies in these various findings that are not easy to interpret. For example, underrepresented students are considerably more likely to offer the attractiveness of the life of an engineering professor as a very important or somewhat important reason for going to graduate school, but far less likely to mention the attractiveness of a career in research or of learning more about a field of study. Asian students seem more oriented towards research and academia, while underrepresented students seem more entrepreneurial. Furthermore, the very small numbers involved make any interpretations risky. The relationships of the academic performance of underrepresented students and of their financial situation to some of their perceptions and choices do present a framework for connecting at least some of the findings, but fail to account for others. Further survey research, involving larger numbers of underrepresented participants, could produce valuable insights.

U.S. Female and Male Students

Some of the salient differences between U.S. female and male engineering seniors are worth reporting because they bear on possible policy interventions to recruit more women into graduate engineering. Female engineering seniors reported more involvement in non-athletic extracurricular activities than did male students. For example in student chapters of professional organizations (82% female vs. 62% male) and tutoring and counseling (33% female vs. 21% male; see Table 14). As in the case of the underrepresented group, the question might be whether the women get too involved in non-academic activities to maximize pursuit of their technical specialization. Yet women reported more encouragement to attend graduate school than males, especially from engineering faculty members (52% female vs. 43% male). A higher percentage of females (40% vs. 32% for males) expect to work towards the doctorate as the highest degree. It should be noted that these are fairly modest differences.

Female and male seniors do not differ markedly in their reasons for attending graduate school in engineering nor in their perceptions of what a doctorate entails. Fewer females than males feel the doctorate would be much more difficult than undergraduate engineering or too much work for too small a future payoff. Generally, the females seem somewhat less motivated by financial considerations than their male counterparts, i.e., by the improved salary prospects contingent on graduate study and the size of the aid package they were offered.

Male and female seniors' reasons for *not* going to graduate school are also rather similar. Those females not going on to graduate work are somewhat more discouraged by the length of time required for the doctorate and the dissertation. Perhaps they are more realistic. Slightly more women (11%)

Table 14
Extracurricular Activities: Female vs. Male

Activity	% Female	% Male
Student government	13.2	9.1
Religious groups	20.1	15.1
Athletics	40.2	55.8
Journalism	4.5	2.6
Music or dance	21.2	10.9
National, state, or local politics (on or off campus)	3.7	3.0
Community service (off campus)	22.8	18.0
Tutoring, counseling or advising	32.8	21.1
Dramatics	2.4	1.7
Fraternity or sorority	25.4	20.5
Honor societies or honoraries	43.7	33.2
Student chapter of engineering or science professional society	82.3	61.9
Other (Please specify)	13.0	11.9
None	2.1	9.2

than men (8%) felt they could not handle graduate work in engineering. While more females appeared to have immediate opportunities to take jobs (more women than men indicated that they had been offered a well-paying job in industry, 36% vs. 29%), they also seemed less likely to take these jobs (fewer felt that it was time they earned a living, 46% vs. 55%).

A mere 1% of each group offered domination of the engineering profession by white males as a *reason* for not going to graduate school. Yet, as shown by responses to another question, 24% of female seniors, as opposed to 7% of male seniors have this *perception* of the profession being dominated by white males, and this difference increases to 37% vs. 14% among graduate students.

In their engineering career aspirations, the female seniors have preferences that are distinct from those of the male students: they prefer sales and marketing (27% vs. 19%), communicating with the public (36% vs. 21%), and making significant practical or useful technical contributions; they are less interested in design and development or in starting their own company (see Tables 15, 16). With regard to starting a company, the difference is greatest; only 20% of women have this aspiration, as against 39% of men. It may be fair to say that the women seniors are more people-oriented and service-oriented than the men. This may be the case because they have accepted this stereotyped view of female roles or it may be their realistic assessment of the occupational roles accessible to them in a world that is dominated by males.

Table 15
Preferred Engineering Activities After Graduation:
Female vs. Male

Activity	% Female	% Male
Research	29.0	29.1
Design and development	59.8	73.7
Production and/or manufacturing	34.6	37.3
Sales and/or marketing	27.1	18.6
Management and administration	55.1	51.8
Communicating with the public	36.2	20.8
Teaching	21.8	20.5
Consulting	39.9	44.4
Other (Please specify)	3.2	2.5
Don't know	1.9	1.7
None	1.9	0.8

Table 16
Career Aspirations: Female vs. Male

Activity	% Female	% Male
To make significant, practical, useful technical contributions in my field	72.5	64.2
To make significant research contributions to furthering knowledge	23.2	22.7
To manage important projects	52.8	58.9
To make a product and sell it	15.2	21.4
To make a comfortable living	73.3	79.9
To contribute to the well-being of my fellow citizens	51.7	44.4
To help improve living conditions for those less fortunate both here and abroad	34.1	28.0
To work for a large company	31.2	27.8
To work for a small company	20.3	19.7
To start my own company	20.3	39.0
To work in federal, state or local government	7.5	8.1
To work for a public interest group	6.4	2.6
To join the Peace Corps	5.3	2.8
To teach at the university level	13.9	11.2
To teach at the secondary school level	8.0	5.7
Other (Please specify)	4.3	2.1

Ph.D.-Bound vs. Nevers (U.S. Undergraduates)

In our sample of U.S. engineering seniors, 452 (19%) planned to go to graduate school immediately, and of these, 148 (33%) aimed at a doctoral degree; 1738 (73 %) said they were not going to graduate school in the fall of 1988 and of these, 328 (17%) said they would never do so. These two groups, those seeking doctorates immediately and those never going to graduate school, represent, in the context of this study, extreme positions and it is therefore especially revealing to compare them.

There are pronounced differences in the undergraduate experiences of the two groups. A higher percentage of the Ph.D.-bound had majors in math or science; the Ph.D.-bound were also more likely to have enjoyed undergraduate courses outside engineering — in math or science or in the social sciences or humanities. A much higher proportion of the "nevers" had negative views of their undergraduate programs; almost half (48%) found undergraduate programs too demanding and 37% too theoretical. The Ph.D.-bound received far more encouragement from faculty and family to go to graduate school than the nevers. About half of the Ph.D.-bound and only 16% of the nevers were involved as undergraduates in tutoring, counseling, or advising other students. The Ph.D.-bound got involved in journalism, music, and dance, and student chapters of professional societies, while the nevers preferred fraternities and sororities; interest in athletics was almost identical.

What does all this mean? The Ph.D.-bound appear to be broader-gauged, more confident academically, better pleased with their undergraduate experience, and more pre-professional in their extracurricular activities. They also receive more reinforcement in their educational aspirations from faculty and family. The fact that 38% of their fathers and 18% of their mothers completed a graduate or professional degree may well explain the greater amount of (non-economic) family support they receive.

The Ph.D.-bound have, not surprisingly, far more positive views of doctoral study than the nevers. A high proportion of the Ph.D.-bound see a doctoral degree as providing access to state-of-the-art equipment and facilities and a strong theoretical basis for an engineering career; they expect doctoral studies to be intellectually stimulating. The nevers are more likely to believe that doctoral study would be much more difficult than undergraduate engineering, too much work for too little payoff, and likely to overqualify them for industry. The career aspirations of these two groups also differ widely, with the Ph.D.-bound more likely to want to work in research, teaching, and design and development, while the nevers are more inclined to work in production, sales, and management.

U.S. Seniors: To Graduate School Immediately vs. Maybe Later

Briefly, we shall compare those who planned to go to graduate school immediately with those who thought they might do so at a later time. What is interesting is that in most respects, the two groups were similar. We found one major difference: those who intended to continue their education immediately received more encouragement from all categories of significant others (faculty, family, friends). The major reasons the "laters" gave for their possible eventual return to school were that they would need to learn more about their engineering field (83%), that their employer would pay for their graduate education (79%), and that their employer would encourage them to get some graduate training (70%). The predominant lack of difference between the responses of the "immediates" and the "laters" suggests that there may be considerable potential for picking up laters later.

Engineering Disciplines

Differences among engineering students majoring in one or another discipline are, for the most part, not very striking. Yet those in civil engineering (CEs) do stand out in certain respects from the rest: Chemical engineers (ChE), electrical engineers (EE) and mechanical engineers (ME).

The CEs were more likely to be satisfied with their undergraduate engineering programs (38% vs. 30% for ChE, 28% for EE, 26% for ME) and fewer CEs found their programs to be too demanding (only 30% of CEs felt burned out, compared to 44% of ChEs). This greater satisfaction does not, however, translate into an especially high proportion of decisions to go to graduate school (EEs are the highest with 22%, and MEs the lowest with 16%) and, indeed, CEs indicate much less frequently than those continuing in other disciplines the intention to obtain a doctorate (19% for CEs, compared to 49% for ChEs).

These disciplinary differences with regard to obtaining doctoral degrees are appropriately reflected in different concerns about the length of time it takes to get a doctoral degree (the CEs are the most easily discouraged of all) and in career aspirations: only 14% of CEs are interested in research careers, compared to 42% of EEs and 32% of ChEs (see Table 17). While EEs and MEs are more interested in teaching and in design and development, CEs and MEs are more inclined towards consulting, and in the case of CEs, communicating with the public. With regard to their desire to serve society and the public interest, CEs are far ahead of the other disciplines: 66% of them hope to do so, compared to 45% of ChEs, 43% of EEs, and 47% of MEs. Clearly, CEs are different.

For any satisfactory interpretation of these findings, it is necessary to put disciplinary differences in the context of occupational opportunities. To a limited extent, the patterns described above may reflect students' perceptions of the difficulty of the different fields of engineering; there may be a tendency for CEs, whose GPAs are typically somewhat lower than those of students in the other fields,* to define this field as easier and to opt for

---

* We are inclined to dismiss the explanation that CE professors are tougher graders.

Table 17
Preferred Engineering Activities After Graduation:
Disciplinary Differences

Activity	Chemical Engineering	Civil Engineering	Electrical and/or Computer Engineering	Mechanical Engineering
Research	32.3	14.4	41.8	30.7
Design and development	57.0	69.7	81.3	74.3
Production and manufacturing	55.6	14.3	32.5	47.9
Sales and/or marketing	23.8	14.4	22.9	19.8
Management and administration	53.4	54.1	45.4	51.0
Communicating with the public	20.4	29.2	21.3	18.6
Teaching	19.2	17.5	27.5	25.9
Consulting	32.5	56.8	35.1	40.8
Other	2.1	3.2	2.2	2.0
Don't know	1.2	1.9	1.2	2.4
None	1.4	1.3	0.8	1.0

the career paths that require less advanced skills. However, this interpretation accounts for only a part of the variation in the attitudes and decisions of students in the various engineering disciplines: the fact that EEs and MEs are more likely to feel that engineering is too much oriented towards military work (21% and 14%, respectively, while only 3% of CEs and 2% of ChEs had this sense) surely reflects what engineers in different disciplines actually do (see Table 18). Finally, the relative lack of interest in Ph.D. study in civil engineering may reflect the nature of that particular professional occupation as well as the relatively low priority it has received vis-a-vis more "high-tech," military-oriented engineering activity in recent years. Given the many physical infrastructure problems the United States faces — crumbling bridges, roads, etc. — it may be time to pay more attention to the advanced engineering requirements of this discipline.

Type of Institution or Department

As noted in the Methodology section our study used a number of institutional variables in order to ascertain whether the type of institution attended makes a difference in the likelihood that engineering students would decide on graduate education and on doctoral as against master's level

Table 18

Impressions of Engineering Profession:
Disciplinary Differences

Impression	Chemical Engineering	Civil Engineering	Electrical and/or Computer Engineering	Mechanical Engineering
Exciting and challenging	74.4	74.5	77.9	74.0
Boring	7.1	9.0	8.2	9.0
Engineers do many useful, practical things to benefit mankind	84.8	86.6	74.1	81.4
Too oriented towards military work	2.6	3.3	20.5	14.0
Too dominated by white males	8.5	10.3	9.8	10.3
Important for the future of the U.S.	72.5	66.1	67.1	71.2
Pays well	78.0	60.0	69.3	69.4
Too closely tied to corporate interests	9.7	3.0	12.0	8.6
Does not project a clear, visible image	9.2	4.9	11.4	12.9
Serves society and the public interest	44.8	66.5	43.0	46.5
Leads to success in business	46.7	39.1	35.5	41.8
Other	5.0	2.5	5.0	3.8

work. Among the variables used were the quality of the department or overall engineering school (QRI and deans' top-20); the quantitative significance of the institution in terms of the production of bachelors degree and doctorate degree holders in engineering (top-20); and its governance, i.e., public vs. private. Two of these variables will be discussed below.

Effects of Quality/Research Intensiveness.   The results of stratifying institutions and departments by quality are rather small in scale: not surprisingly, better schools have better students, and more surprisingly, in many other respects, the students at the better schools responded in the same way as students at the less good schools. Even with regard to post-baccalaureate plans the differences are not very great: the percentages of seniors in

QRI-1, QRI-2, and QRI-3 departments* (roughly in order of decreasing quality) who plan to attend graduate school are 83%, 78%, and 71% respectively.

With regard to reasons for attending or not attending graduate engineering school, students in QRI-1 institutions more frequently offered their academic ability as a very important reason (47% for QRI-1, 32% for QRI-2, and 35% for QRI-3); those *not* planning to go to graduate school who attended departments or institutions distinguished by their quality tended more frequently to offer having had enough of school as their leading reason than did students at institutions of lesser quality. It seems possible that the academic pressure on students in high-quality departments backfires when it comes to recruiting graduate students. Not surprisingly, QRI-1 students less frequently respond that they have no attractive full-time job offers and that they have been offered a good financial package. By contrast, no conspicuous differences are found among students in departments of different quality when it comes to plans for the possible postponement of graduate education until a later time and the reasons for it. Finally, no significant differences emerge with regard to views of the engineering profession, career aspirations, or preferred post-educational engineering activities.

Effects of Governance. Differences resulting from governance, i.e., between public and private institutions, are partly confounded by quality differences: 37% of the departments in public institutions are in the QRI-1 (highest quality) category, compared with 22% of departments in private institutions. Therefore, it is not possible to attribute differences in the responses of students unequivocally to governance. Rather similar percentages of students (22% in private institutions, compared to 17% in public ones) plan to go on. Yet students in public institutions are far more likely to apply to the institution from which they will receive their undergraduate degree (73% vs. 43% for private) — a large difference that may reflect quality differences or economic factors or both. More significantly, students in private institutions are substantially more inclined to work for the doctorate as their highest degree (41% vs. 29%). The reason is elusive: it may be partly a result of the disciplines of survey respondents; there is a somewhat higher proportion of chemical engineers among the respondents from private institutions and a higher proportion of civil engineers among those in public ones, although the differences are too small to account for the 12% difference in the Ph.D. goal.

---

* In defining our three QRI groups, we were uncertain of the quality of the clearly less research-intensive QRI-3 group compared with QRI-1 and QRI-2. Some quality-related findings do indicate that the groups decrease in quality in the order QRI-1, QRI-2, and QRI-3.

Students in private institutions seem more sensitive to economic considerations, like having incurred large debts for undergraduate education. Since
there are not very pronounced differences between the income distribu-
tions of students in public and private institutions, a surprising result, the
matter of debt may be directly related to the cost of undergraduate private
education *per se*. In any event, among students in private institutions who
decided against graduate education, 36% (vs. 24% in public schools) offered as a primary reason their large debts; 48% of them (vs. 41%) said they
planned to work before going to graduate school. In private institutions, 33%
of students said they were not going to graduate school because they had
been offered a well-paying job in industry, compared with 27% in public institutions.

We had hypothesized, in stratifying our survey by type of institution, that
the atmosphere or academic culture in different kinds of schools might be
more or less conducive to decisions to continue engineering education beyond the bachelors degree. Our findings suggest that this is not the case.*
It is the academic performance of individuals that seems to make the key
difference in defining their sense of what they are capable of with regard
to advanced education and its occupational outcomes, coupled with the assessment of students in different economic situations of the economic implications of graduate school. The quality or type of institution they choose
for their first degree appears to make far less difference.

---

* Perhaps it might have been if we had tried to take much narrower quality cuts — say, com-
paring students in the "top-20" schools with those in several other strata. However, what little
analysis we did of students in the "top-20" schools vs. those in all other schools does not
bear this out.

Findings: Graduate Students

The demographic composition of the population of engineering graduate student respondents, although rather similar to that of the undergraduate population in many respects, differs in one major way. The big difference is in the proportion of foreign students: 43% of graduate students are not U.S. citizens, compared to 9% of undergraduates. The percentage of females is even lower than among the undergraduates (11% vs. 17%); the percentage of U.S. students from all minority groups is quite similar (11% vs. 12%), but if we consider the underrepresented group, blacks and Hispanics and Native Americans separately, they total only 3%, compared to 6% among undergraduates (see Table 1).

The questionnaire administered to graduate students has a somewhat different emphasis from the one given to seniors. It contains sets of questions on the circumstances of the decision to go to graduate school; and the decision to pursue a master's or a Ph.D. degree. Like the questionnaire for seniors, this one includes questions about their views of the engineering profession, the kinds of activities they would prefer to pursue, and their career goals, as well as about their personal and family backgrounds. In the next two sections, we shall present overall findings and findings related to academic ability; in only a few instances we shall compare U.S. and foreign students. In later sections, we shall deal with differences related to citizenship, gender, racial/ethnic background, and the type of institutions where the students are enrolled.

Decisions to Go to Graduate School

Some 43% of students who went beyond the bachelors degree decided to do so in the last half of their undergraduate education, and almost one-fifth (19%) decided after completing their undergraduate education and working full-time for a year or more (see Table 19). This pattern of making relatively late decisions to go to graduate school is even more pronounced if the students' academic performance is taken into account: fully 50% of those with GPAs in the highest quartile decide during the last half of their undergraduate education, while 25% of those with GPAs in the lowest quartile wait until they have worked a year or more. Overall, 17% make their decision before they even begin their undergraduate studies, and among the top students the proportion is almost one-fifth. It might be worth exploring further whether early and late decisions to go to graduate school are related to confidence in the ability to perform effectively there. Is it that the somewhat less able students require more time to gain the confidence to obtain graduate education or do they, at a certain point, see graduate education as the last, best hope in their quest for interesting and well-paying jobs?

Table 19
Time of Deciding on Graduate Degree

Time of Decision	Frequency	%
Before I began my undergraduate education	406	16.7
During the first half of my undergraduate education	195	8.0
During the last half of my undergraduate education	1,036	42.5
Just after completing my undergraduate education	173	7.1
After completing my undergraduate education and working full-time for a year or more	473	19.4
After completing my undergraduate education and not being able to get the job I wanted	122	5.0
Other	31	1.3
Total	**2,436**	**100.0**

If we compare foreign and U.S. graduate students with regard to the timing of the decision to go to graduate school, the patterns vary quite noticeably. Among foreign graduate students, as many as a quarter (25%) decide on graduate education before beginning their undergraduate studies (vs. 11% for U.S. students), while only 35% make their decision during the last half of their undergraduate education. This seems to be one indication, and there will be more discussed below, that foreign students and their families tend to be more strongly persuaded early on that graduate education is desirable.

The three groups that generally played the largest part in encouraging the decision to go to graduate school were the students' families (reported by 72%), engineering faculty at their undergraduate schools (69%), and their friends (57%) (see Table 20). However, among the best students, 79% reported encouragement by faculty, while among the lowest GPA group, only 58% did so. Curiously, encouragement by families and friends was quite unrelated to academic performance. And though, overall, few graduate students (17%) reported aggressive recruitment by universities (compared to

51% who reported aggressive recruitment by industry; see Table 21), 57% of the top students thought universities showed an interest in individuals, compared to 44% of the lowest quartile students.

Table 20

Sources of Encouragement of Graduate Education:
Graduate Respondents

Source	Overall Encouraged Me		Overall Discouraged Me		Neutral or Not Applicable	
	Frequency	%	Frequency	%	Frequency	%
Engineering faculty members at undergrad. school	1,665	68.5	47	1.9	718	29.5
Non-engineering faculty at undergrad. school	385	15.9	42	1.7	1,990	82.3
Graduate teaching assistants at undergrad. school	493	20.4	65	2.7	1,858	76.9
My family	1,764	72.3	110	4.5	567	23.2
My fellow students	1,088	45.1	168	7.0	1,154	47.9
My friends	1,377	56.8	154	6.4	894	36.9
My spouse	602	25.2	53	2.2	1,732	72.6
My employer	428	18.0	133	5.6	1,816	76.4
Other	65	22.3	11	3.8	215	73.9

Table 21

Perceptions of Recruitment Practices:
Universities vs. Industry

Preception	Agree		Disagree		Don't Know	
	Frequency	%	Frequency	%	Frequency	%
Industry recruits aggressively	1,198	50.8	424	18.0	736	31.2
Universities recruit aggressively	404	17.1	1,298	55.1	655	27.8
Industry shows interest in individuals	1,214	51.5	396	16.8	745	31.6
Universities show interest in individuals	1,167	49.6	636	27.0	550	23.4
Industry provides funds for visits	1,452	61.7	165	7.0	736	31.3
Universities provide funds for visits	456	19.3	984	41.7	916	38.9
Industry provides needed information	1,311	55.7	329	14.0	711	30.2
Universities provide needed information	1,716	73.2	252	10.8	375	16.0

As among the engineering seniors, the predominant number of graduate students gave idealistic reasons for their decision to pursue advanced degrees: 73% said a very important reason was the interesting and challenging nature of the subject matter and 68% said they wanted to know more about their field (see Table 22). On a more practical level, 59% saw graduate study as a very important prerequisite to doing the work that interests them and 45% saw it as very important for the research careers they hoped to pursue. As many as 46% turned down full-time job offers or gave up a full-time job to attend graduate school, and 76% responded that the lack of an attractive job offer played no important role in their decision to attend graduate school.

Doing well academically is certainly a factor in the students' decisions to go on to graduate school: 43% said doing well academically was very

Table 22
Importance of Reasons for Attending Graduate School:
Graduate Respondents

Reason	Very Important		Somewhat Important		Not Important/ Not Applicable	
	Frequency	%	Frequency	%	Frequency	%
The subject matter is interesting and challenging	1,772	72.9	600	24.7	58	2.4
I do well in academic work	1,057	43.4	1,093	44.9	284	11.7
I need a graduate degree to do the work that interests me	1,434	58.9	708	29.1	294	12.1
I want to pursue a career in research	1,087	44.5	792	32.5	561	23.0
I want to improve my future salary prospects	813	33.4	1,088	44.7	531	21.8
I had no attractive full-time job offers	219	9.0	357	14.7	1,849	76.2
I was not ready to go to work	209	8.6	509	21.0	1,707	70.4
I want to learn more in depth about my field	1,651	67.7	648	26.6	139	5.7
In general, my family regards advanced degrees highly	601	24.6	719	29.5	1,119	45.9
Many of my friends have, or intend to get, advanced degrees	203	8.3	543	22.3	1,687	69.3
Other	200	53.6	21	5.6	152	40.8

important and another 45% that it was somewhat important. For the students in the highest GPA quartile, a sense of academic competence played a much stronger role; 65% identified it as a very important factor in their decisions, while only 30% of the lowest GPA quartile students did so.

We have some interesting findings about the way in which graduate students perceive the financial circumstances of their decisions to go to graduate school. Almost half (49%) identified a good financial aid package as a very important factor, though 27% considered a good financial aid package to be not important or not applicable. Almost a quarter of them (24%) gave high importance to the judgment that the long-term economic rewards of graduate study would eventually more than balance the immediate costs. Only 12% responded that full or partial funding of their graduate studies by the company for which they worked was either a very important or a somewhat important factor in their decision to attend graduate school full-time.

The importance of a good financial aid package figured even more prominently in the decisions of the best students than in the population as a whole: in the highest-quartile GPA group, 59% said this was a very important factor (compared to 41% of the lowest-quartile GPA students). By contrast, the students with the lowest GPAs were slightly more likely to emphasize the importance of the long-term economic benefits of graduate study.

Graduate Students — M.S. vs. Ph.D.

At the time of our survey, 54% of our graduate student respondents were pursuing master's degrees and 46% were pursuing doctoral degrees. (It should be recalled here that doctoral students were oversampled in our survey because of special policy concerns about the supply of engineers with Ph.D. degrees.) More significantly, 69% responded that the doctoral degree was the highest degree they expected to obtain. If these data are analyzed in terms of the academic performance of the graduate students, as indicated by undergraduate GPAs, 57% of those in the highest GPA quartile were pursuing Ph.D.s, and a startlingly high 78% of this group were headed for eventual doctorates. To be sure, only 19% of undergraduates opt for graduate school in the first place, but once enrolled, high proportions of those surveyed plan to go all the way and among the best students, the proportion is very high indeed. Our data indicate that, given the large supply of undergraduates, attracting relatively small additional percentages of them into graduate education could appreciably increase the flow of Ph.D.s into the pipelines mentioned above. However, this result needs further verification because it may be attributable to overrepresentation of doctoral students, which also makes it difficult to estimate the increase of flow with any precision.

The reasons offered by all students (U.S. and foreign combined) for pursuing doctoral degrees are perhaps somewhat more practical than those offered by graduate students for being in graduate school in the first place. Their most frequent response was that they desired to learn more about a field (82%), a reason that may have both intellectual and practical components; and the next most frequently offered reasons were to improve career opportunities (72%), and to prepare for a position in research (71%); prestige and improved earning power were much lower down on the list (see Table 23). Differences according to GPA were not very conspicuous.

Table 23
Reasons for Obtaining Doctorate

Reason	Frequency	%
I want to learn more in depth about my chosen field	1,273	81.7
A doctorate will improve my career opportunities in general	1,128	72.4
I want a position doing research	1,104	70.9
I need it as a credential for a faculty position	811	52.1
The prestige of a doctoral degree will increase my chances of advancement to high levels in management or administration	522	33.5
My future increased earning power will outweigh the time and money invested in getting a doctorate	430	27.6
I'm not ready to get a job	179	11.5
Other (Please specify)	77	4.9

The kinds of reasons offered by those who did not want to pursue doctoral degrees do not seem easily susceptible to policy intervention: these graduate students said they did not want to become professors (63%); they found the time required to get a doctoral degree too long (58%); they deemed such a degree not cost effective (51%); and they thought a doctorate was too theoretical and research oriented (46%; see Table 24). However, one-third or 33% said that they lacked the financial resources for doctoral study. Again, the variable of academic ability (as measured by the GPA proxy) did not produce very different results. In any event, economic considerations do play some part in the decisions to obtain a doctoral degree and may provide valuable leverage for policy intervention, but career preferences appear at least as important.

Given the strong interest in the relative dearth of U.S. doctorate recipients in engineering, it is important to find out why the U.S. students in our sample decided to take the doctoral as against the master's degree route. Breaking down the graduate student population into these two groups is parallel to breaking the undergraduate group into the Ph.D.-bound and the Nevers, as discussed previously.

Table 24

Reasons for Not Obtaining Doctorate

Reason	Frequency	%
I am not interested in a job as a professor	270	63.1
It takes too long to get a doctorate	250	58.4
The time and money invested in getting a doctorate will be greater than the benefits of having one	220	51.4
A doctorate will not help me to do the work I want to do	216	50.5
A doctorate is too theoretical and research-oriented for me	198	46.3
I don't have the financial resources	139	32.5
A doctorate will overqualify me for jobs in industry	132	30.8
I don't think I can handle doctoral level work	71	16.6
Other (Please specify)	53	12.4
I have been discouraged for academic reasons from going for a doctorate	41	9.6
I'm obligated to return to my job after my master's	26	6.1

Among the U.S. graduate students in our sample, 799 or two-thirds indicated their intention to pursue a doctoral degree as their highest degree and 393 or one-third intended to stop at a master's degree. Not too surprisingly, the undergraduate GPAs of the Ph.D.-bound were higher than those of the M.S.-bound; overall, the Ph.D.-bound had GPAs of 3.50 compared to 3.32 for the M.S.-bound, and in engineering courses, 3.57 for the Ph.D.-bound vs. 3.38 for the M.S.-bound. As in the case of the undergraduate Ph.D-bound, this subgroup at the graduate level consisted of stronger students and exhibited some of the confidence in their decisions and aspirations that more successful students are likely to have. Thus, 48% of them, as against 29% of M.S.-bound students, offered as a reason for attending graduate school that they did well in academic work, and 20% of the M.S.-bound as against 7% of the Ph.D.-bound wound up in graduate school at least in part because they had no attractive full-time job offers.

A number of other differences should be noted here. The U.S. Ph.D.-bound graduate students decided at an earlier time than the U.S. M.S.-bound to obtain graduate degrees (14% of them did so before they began their undergraduate education, compared to 6% of the M.S.-bound). And once in graduate school, the Ph.D.-bound explained their reasons for being there in rather different directions from the M.S.-bound: they were much more likely to say they needed a graduate degree to do the work that interests them (71% vs. 43%), that the subject matter of engineering is interesting and challenging (82% vs. 70%) and, by a lopsided margin of 59% to 14%, that they want to pursue a career in research. (Correspondingly, 53% of the M.S.-bound see a research career as a reason that is not important or not applicable, as against 11% of the Ph.D.-bound.)

The same kinds of reasons that lead U.S. students into Ph.D. as against master's programs are, not surprisingly, reflected in their views about the engineering profession and the kind of career they would prefer. The M.S.-bound are more likely to see engineering as a stepping stone to success in business (31% vs. 25% of Ph.D.-bound) and they are more likely to want to be involved in design and development (81% vs. 65% for the Ph.D.-bound), in production and manufacturing (30% vs. 12%), and in making a product and selling it (53% vs. 31%). By contrast, the Ph.D.-bound are much more interested in research (90% vs. 40%), in teaching generally (68% vs. 20%) and university teaching (60% vs. 7%), and in consulting (61% vs. 47%; see Table 25).

Table 25
Preferred Engineering Activities:
M.S.-Bound vs. Ph.D.-Bound

Activity	% M.S.	% Ph.D.
Research	39.7	89.6
Design and development	81.4	65.0
Production and/or manufacturing	29.5	11.7
Sales and/or marketing	10.2	3.9
Management and administration	38.9	26.3
Communicating with the public	15.5	18.8
Teaching	20.1	68.2
Consulting	46.6	60.8
Other (Please specify)	3.1	2.0
Don't know	1.3	0.8

As we noted in the case of the two polar U.S. undergraduate groups, the Ph.D.-bound and the nevers, the Ph.D.- and M.S.-bound groups of graduate students are also very different people. Here again, the highest level of parents' education is higher for the Ph.D.-bound: 38% of their fathers have completed graduate or professional degrees, as have 19% of their mothers, compared to 28% of the fathers of M.S.-bound students and 9% of their mothers. Although they are all graduate students, the ones who are heading for Ph.D.'s are far more committed to understanding and advancing the substance of engineering, while the M.S.-bound are interested in applying what they learn in an industrial and/or commercial context. This suggests that care would need to be exercised in encouraging those M.S.-bound U.S. graduate students who might make suitable material for the academic pipeline or the high-level research pipeline into industry.

Images of Engineering and Career Aspirations

Although only a small minority of engineering undergraduates choose to go to graduate school, the views of both U.S. and foreign graduate students about the nature of the engineering profession, are remarkably similar to the views of undergraduates (see Table 26). Almost as high a percentage of graduate students as undergraduates think engineering provides many useful and beneficial things to mankind (79% for graduate students vs. 87% for undergraduates) and 80% of graduate students as against 75% of undergraduates think engineering is exciting and challenging. Perhaps an important difference is noticeable in the proportions of the two groups that think engineering pays well: 55% of the graduate students and as many as 68% of the undergraduates. Also, only 26% of the graduate students and 40% of the undergraduates think engineering leads to success in business. Those self-selected to graduate study seem somewhat less optimistic, then, about the economic rewards of being engineers.*

Table 26
Impressions of Engineering Profession:
Graduate Respondents

Impression	Frequency	%**
Exciting and challenging	1,941	79.9
Boring	112	4.6
Engineers do many useful, practical things to benefit mankind	1,905	78.5
Too oriented towards military work	427	17.6
Too dominated by white males	321	13.2
Important for the future of the United States	1,424	58.6
Pays well	1,344	55.4
Too closely tied to corporate interests	322	13.3
Does not project a clear, visible image	311	12.8
Serves society and the public interest	1,130	46.5
Leads to success in business	638	26.3
Other	117	4.8

** Percent based on those responding. Multiple responses allowed.

When it comes to attractive future activities, however, there are more significant differences between graduate students and undergraduates. Although both groups are very interested in design and development (70% for graduate students vs. 71% for undergraduates), the activity preferred by graduate students is research (72% vs. 29% for undergraduates; see Table 27). Also, 53% of graduate students can see themselves interested in teaching compared to 21% of undergraduates. Again, we have two groups with quite different ideas of what they hope to be doing.

---

* Differences could be attributable to the very different proportions of U.S. versus foreign students in the graduate versus undergraduate populations.

Table 27

Preferred Engineering Activities After Graduate Degree

Activity	Frequency	%
Research	1,760	72.2
Design and development	1,694	69.5
Production and/or manufacturing	473	19.4
Sales and/or marketing	164	6.7
Management and administration	693	28.4
Communicating with the public	325	13.3
Teaching	1,292	53.0
Consulting	1,277	52.4
Other	49	2.0
Don't know	28	1.1
None	13	0.5

Finally, there is the matter of career aspirations. Again, there are some notable differences between undergraduates and graduate students. While the leading aspiration of engineering undergraduates is to make a comfortable living (77% have this goal), only 61% of graduate students have this aspiration (see Table 28). This disparity may be related to another one: among graduate students, 46% overall hope to teach at the university level, but only 12% of undergraduates do so. Both groups put a lot of stock in making significant practical, useful contributions to their field (73% of graduate students and 65% of undergraduates), but there is a striking difference in the importance they attribute to research: 57% of graduate students, and 80% of those graduate students with the highest undergraduate GPAs, hope to make significant contributions to furthering knowledge through research, while a mere 23% of all undergraduates do so, and even among the top-quartile GPA undergraduates the percentage is only 42%.

Our findings show quite clearly that both undergraduates and graduate students are keenly interested in the application of engineering knowledge, but the latter group is more committed to the actual advancement of engineering knowledge itself.

Foreign vs. U.S. Citizenship

The high proportion of foreign students in U.S. graduate engineering programs permits comparisons between their perceptions and expectations and those of U.S. graduate students. Such comparisons may illuminate the decisions of the U.S. students and reveal ways of influencing the nature of these decisions. It is especially important to try to identify differences in the various reasons indicated by foreign and U.S. students for attending

Table 28
Career Aspirations: Graduate Respondents

Aspiration	Frequency	%*
To make significant, practical, useful technical contributions in my field	1,757	72.5
To make significant research contributions to furthering knowledge	1,377	56.8
To manage important projects	924	38.1
To make a product and sell it	345	14.2
To make a comfortable living	1,482	61.2
To contribute to the well-being of my fellow citizens	948	39.1
To help improve living conditions for those less fortunate both here and abroad	724	29.9
To work for a large company	558	23.0
To work for a small company	301	12.4
To start my own company	620	25.6
To work in federal, state or local government	129	5.3
To work for a public interest group	98	4.0
To join the Peace Corps	43	1.8
To teach at the university level	1,104	45.6
To teach at the secondary school level	80	3.3
Other	69	2.8

* Percent based on those responding. Multiple responses allowed.

graduate school. What makes graduate study worthwhile to each group? And contrarily, why is graduate study an unattractive option?

Foreign students start with some handicaps that may be reflected in GPAs. (All GPAs reported for graduate students are those they obtained as undergraduates. Since most foreign graduate students obtained their undergraduate training in their home countries, the comparison of their GPAs with those of U.S. students cannot be made with great confidence.) The mean overall GPA of U.S. students is higher than that of foreign students (3.43 vs. 3.27) because of higher GPAs in courses both *outside* of engineering, science, and math (3.46 vs. 3.22), and *in* engineering courses (3.58 vs. 3.40). The percentage of foreign students in the three highest quartile groups is in the range of 33% to 35%, whereas for the lowest quartile group it is 59%. Presumably, some of these GPA differences reflect the language difficulties of foreign students which also emerge in differences in Graduate Record Examination (GRE) scores. On the GRE, U.S. students scored higher on the verbal component (589 vs. 466) and the analytical component (684 vs. 595), while foreign students did better on the quantitative portion (739 vs. 719). Foreign students reported a mean approximate TOEFL score of 590.

A significant handicap for foreign students may be the greater likelihood that they are not only going to school in a new country but also attending a new school to which they need to become accustomed. While 42% of U.S. students were attending the same institution from which they received their bachelors degree in engineering or science, only 12% of foreign students (most of whom did undergraduate study in their home countries) were attending the same institution. As our survey results confirm, for a good many U.S. graduate students, it is much easier to obtain all kinds of practical information about the graduate school they are attending and to get advice from faculty because they simply continue at the place where they have received their bachelors degree. Foreign students have to rely far more heavily on college catalogs, the reputation of the institution, and the reports of friends (see Table 29).

Table 29
Sources of Information About Graduate School:
Foreign Students with U.S. B.S. vs. Foreign Country B.S.

Source	Foreign	U.S.
Posters on bulletin boards	8.2	10.5
Faculty members at my previous institution	25.0	22.7
The engineering graduate school contacted me directly	1.0	4.1
Workshop or seminar	1.7	1.2
By reputation or from friends	53.2	36.1
College catalogs or guides (Peterson's, Barron's, etc.)	50.2	19.8
I received an undergraduate degree here	0.2	41.3
Other (Please specify)	11.2	8.7

So much for the handicaps of foreign students that were identified in this study; in a previous study[2] the communications handicaps of foreign students emerged very clearly. Yet we shall see below that foreign students are not deterred by such handicaps in seeking the highest degree that engineering schools offer.

Foreign and U.S. students do not differ noticeably in the reasons they offered for attending graduate school (see Figure 1). An important difference lies, however, in the sources of non-economic support or encouragement. Consistent with the responses of foreign undergraduates, we find considerably more foreign graduate students reporting not only that their family regards advanced degrees highly (68% for foreign students vs. 43% for U.S. students), but also that family and friends provided important encouragement. By contrast, U.S. students more often received encouragement from faculty members, graduate teaching assistants, and employers. Family encouragement for foreign students may be related to the fact that engineering tends to run in their families to a greater degree, as our survey results indicate. This may be one reason for an earlier awareness of what engineering is about and for earlier decisions about graduate study in engineering, as mentioned above.

Figure 1
Very Important Reasons for Attending Graduate School

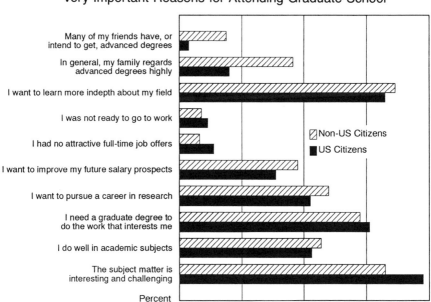

There appear to be no important differences in the extent to which U.S. and foreign students were influenced by academic and career-related reasons or by economic factors in their decision to go to graduate school. U.S. students are somewhat more likely than foreign students to be offered financial aid to attend their first year of full-time graduate study, a finding that confirms Barber and Morgan's 1987 study that showed a disposition by chairpersons and faculty to make funds somewhat more available to U.S. students (see Figure 2). The mean monthly amount of U.S. students' awards is higher ($791 for U.S. students vs. $655 for foreign students, excluding tuition), but by and large, many U.S. students are accustomed to a higher standard of living than are most foreign students. Roughly identical percentages of both groups report turning down full-time job offers or giving up full-time jobs to attend their current graduate programs (see Figure 3); the mean annual salary given up or turned down was $31,000 for U.S. students, compared to $17,000 for foreign students. (This low figure may be due to lower salaries in the foreign students' home country, as well as to currency conversion uncertainties.)

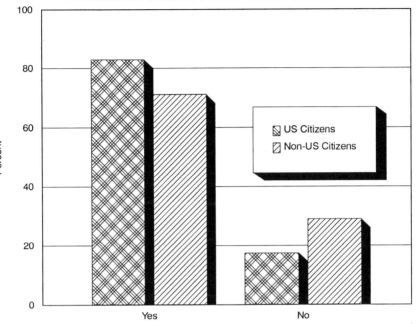

Figure 2
Financial Aid Offers to Attend Graduate School

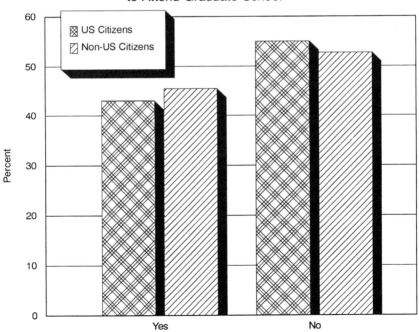

Figure 3
Full-time Jobs Given Up or Offers Turned Down
to Attend Graduate School

In spite of the fact that foreign students embark on graduate studies with the handicaps noted above, a considerably higher percentage of the foreign graduate students surveyed, as many as 81%, expected to obtain doctoral degrees, compared with 59% of U.S. respondents. The foreign students were only slightly more interested than U.S. students in teaching but, perhaps surprisingly, also more interested in production and/or manufacturing. U.S. graduate students more frequently expressed interests in design and development, management and administration, and communicating with the public. The level of interest in research was identical (see Figure 4). U.S. engineering graduate students showed more interest in making a comfortable living after obtaining their degree (70% vs. 51% for foreign graduate students), but they also were more disposed to careers that make contributions to the well-being of their fellow citizens (46% vs. 31%) and making significant, practical contributions to their field (76% vs. 69%).*

Figure 4
Preferred Engineering Activities After Graduation

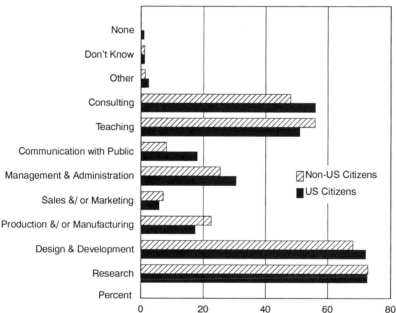

The views of the engineering profession of foreign and U.S. students differed to some extent. For the most part, the perceptions of foreign students were more positive than those of U.S. students, except that twice as high a percentage of foreign students (17% vs. 8%) saw the engineering profes-

---

* Those foreign graduate students who got their bachelors degrees in the U.S. tended to respond to questions concerning career aspirations, views of the profession, etc. similarly to U.S. graduate students, suggesting some acculturation at the undergraduate level.

sion as too dominated by white males. It may be significant also that only 44% of foreign graduate students, compared to 66% of U.S. graduate students, thought that engineering pays well, reflecting perhaps their expectation of working in the academic sphere.

As noted above, our data indicate that the far stronger inclination of foreign students to obtain Ph.D. degrees cannot be explained by higher academic ability. Perhaps it can be accounted for by the kinds of activities they aspire to: in some respects, but not all, the activities preferred by foreign students are more academic and those of U.S. students more practical. However, another explanation can be suggested for the pervasive quest by foreign students of doctoral degrees. It may have its roots in the perception overseas that this is the only degree worth obtaining, i.e., the only degree that carries sufficient prestige to open any desirable occupational doors, even permitting doctorate recipients to migrate to the United States. Foreign students may realize that they have fewer options. For one thing, if they do not remain in school they cannot continue to hold temporary visas. Furthermore, to compete effectively not only for academic positions at home and abroad, but also for positions in American industry, they may assume that they must have doctoral degrees. To find out if this assumption is justified, it would be necessary to learn more about actual competition for industrial jobs between U.S. and foreign engineers.

If foreign engineering students are, indeed, much more likely than U.S. engineering students to assume that only an advanced degree, indeed a Ph.D. degree, will enable them to obtain access to desirable jobs and/or remain in the United States, this may mean that only a complicated set of incentives — of sticks and carrots — is likely to induce more U.S. engineering undergraduates to continue their education. To put it most dramatically, U.S. students would have to be confronted with a situation more like the one facing foreign students. It would be necessary to modify the value of a bachelors degree by changing, for certain positions in industry, the degree requirements that enable many U.S. citizens with B.S. degrees in engineering to pursue the kinds of activities that both interest them and pay well. So long as there is wide acceptance of the norm that, for many U.S. citizens, a first engineering degree *should* produce satisfactory occupational results, as well as abundant evidence that it *does* produce such results, who can expect a considerable number of engineering seniors to reason that a graduate degree is worth the effort and the income foregone? We shall revert to this issue in our conclusions.

For a minority, of course, the subject matter is sufficiently interesting and the prospect of graduate education is not daunting. Among these there are significant numbers who understand that they need a graduate degree to do the kind of work that interests them, including high-level research and university teaching, and that an advanced degree might well have long-term payoffs in terms of higher salaries. But the present situation facing U.S. engineering undergraduates, as well as their families, is very different from that of foreign undergraduates: it is, accordingly, not generally conducive to self-selection into graduate education.

U.S. Racial/Ethnic Groups: Underrepresented (Black, Hispanic, American Indian), Asian-American, and White

If the number of responses from U.S. undergraduates who were members of the underrepresented group or Asian-Americans was small, among U.S. graduate students the number of these respondents was even smaller. The total number of non-white respondents was 130: 38 from the underrepresented group (8 blacks, 26 Hispanics, and 4 American Indians) and 92 from Asians. Especially with regard to the underrepresented group, we cannot claim that our data are significant, yet, once again, it seems useful to provide some findings that may be at least of anecdotal interest.

Several such findings stand out. One is that, like foreign students, underrepresented graduate students are less likely to be attending the institution from which they received their undergraduate degrees (32% of them, as compared to 43% of white graduate students). This means more difficult adjustment problems for them. Another interesting finding is that a higher proportion of the underrepresented group than of the white and Asian-American groups indicated that the fact that the long-term economic rewards more than balance the immediate costs was either a very or somewhat important factor in their decision to attend graduate school. This applied also to the decision to obtain a doctoral degree: more members of the underrepresented group gave as a reason that that their future increased earning power would outweigh the time and money invested in getting a doctorate. Yet another finding worth reporting is that Asian-American graduate students said their families regard advanced degrees highly and that they thought the prestige of a doctoral degree would increase their chances of advancement to high levels in management or administration. A fourth finding is that underrepresented students seem more inclined than white students to teach at the university level (53% vs. 41%). Finally, it appears that underrepresented graduate students are twice as likely as white students to see engineering as too dominated by white males (32% vs. 16% for white students) and too closely tied to corporate interests (27% for underrepresented as against 11% for white).

However shaky the empirical basis of these findings may be, they suggest that engineering graduate students who are members of both the underrepresented and Asian groups see aspects of their situation differently from white graduate students, and in some respects in ways similar to the situation of foreign students. Some of these U.S. graduate students seem similarly inclined to believe that it is only by obtaining advanced degrees that they can compete effectively for well-regarded occupational rewards in the engineering field. Given the great importance of understanding the motivation and decisions of underrepresented students, it would be highly worthwhile to obtain statistically reliable data about their perceptions of the conditions that make different academic degrees in engineering and different types of engineering careers attractive.

U.S. Female and Male Graduate Students

At the graduate level, differences in responses between U.S. female and male students are generally very small. A slightly higher proportion of female respondents belonged to minority groups (15% vs. 10% for males) and 48% of them, compared to 35% for males, had fathers or stepfathers who were engineers. This last finding is similar to that of the occupational background of the fathers of female engineering undergraduates and suggests that recruiters might find good prospects among the daughters of engineers.

U.S. women are not quite as likely as men to anticipate long-term rewards from graduate education (59% of the women did so, compared to 62% of men), nor do they think a doctoral degree will improve their career chances in general (59% vs. 70% for males). The women are either more pessimistic or perhaps more realistic than foreign students or members of non-white racial/ethnic groups about the likelihood that an advanced degree will have long-term benefits.

The views of women graduate students about the engineering profession were somewhat less positive than those of male students. Not surprisingly, a substantially higher percentage of women (37% vs. 14%) felt that engineering is too dominated by white males. While female graduate students were less likely to feel that the engineering profession projects a clear image, they attributed to it less disposition to serve society and the public interest than did male students.

The preferred activities of the female graduate students have considerable continuity with those of female undergraduates, and both groups diverge from male patterns on some responses. Women graduate students showed somewhat less interest in design and development than men; they were much more interested in communicating with the public (26% vs. 16%). Again, they were much less likely to want to start their own company (16% vs. 28%). Only 61% of women, compared to 71% of men, responded that they hoped to make a comfortable living (see Table 30).

Table 30
Career Aspirations: Graduate Respondents

Aspiration	% Female	% Male
To make significant, practical, useful technical contributions in my field	77.4	75.3
To make significant research contributions to furthering knowledge	59.1	54.7
To manage important projects	37.0	39.9
To make a product and sell it	10.5	15.8
To make a comfortable living	60.8	70.9
To contribute to the well-being of my fellow citizens	45.9	45.8
To help improve living conditions for those less fortunate both here and abroad	29.8	27.2
To work for a large company	23.2	19.4
To work for a small company	14.4	16.2
To start my own company	15.5	27.6
To work in federal, state or local government	6.6	4.8
To work for a public interest group	3.3	2.7
To join the Peace Corps	4.4	1.3
To teach at the university level	37.6	41.6
To teach at the secondary school level	7.2	3.1

Whether because of or in spite of the fact that engineering is perhaps *the* most nontraditional field of study for women, the female graduate students tended to see themselves winding up in a somewhat different "place" than the males. Becoming an engineer and eventually a candidate for an advanced degree in engineering does not necessarily mean having the same reasons or expectations as those of male fellow students. However, for the most part, the similarities between female and male career aspirations and views of the engineering profession appear to outweigh the differences. Our study only scratches the surface of women students' special perceptions of their comparative strengths and opportunities. Further research would surely pay off for those concerned about recruiting more women into graduate engineering programs and more women faculty for engineering schools.

Ph.D. Candidates — Academia-Bound vs. Others

Another interesting perspective on the U.S. engineering graduate students who plan to obtain Ph.D.s is to compare those who aspire to academic careers — teaching at the university level — with those who do not. Both groups are overwhelmingly disposed to do research; the non-academic group is slightly more concerned about a comfortable living (70% vs. 63%) and about making practical, useful contributions to society (83% vs. 75%). Where fairly significant differences show up are in the extent to which the academic group might be defined as more idealistic than the non-academic one. Those headed for university teaching are a good deal more likely to consider the engineering profession as too oriented to military work (28% academic vs. 17% nonacademic), too closely tied to corporate interests (18% vs. 9%), and too dominated by white males (21% vs. 15%; see Table 31). The academia-bound group is more inclined to help improve living conditions for those less fortunate, both in the United States and abroad (31% vs. 21% nonacademic) and to contribute to the well-being of fellow citizens (49% vs. 40%).

Table 31
Impressions of Engineering Profession:
Academic Career-Bound vs. Others

Impression	Academic	Non-Academic
Exciting and challenging	85.4	80.8
Boring	6.6	4.3
Engineers do many useful, practical things to benefit mankind	83.2	79.9
Too oriented towards military work	27.6	16.7
Too dominated by white males	21.0	15.2
Important for the future of the United States	86.2	80.8
Pays well	65.2	61.9
Too closely tied to corporate interests	17.8	8.7
Does not project a clear, visible image	20.8	20.7
Serves society and the public interest	52.4	48.6
Leads to success in business	22.3	27.9
Other	4.5	5.9

When the group of academia-bound U.S. students is compared to the foreign academia-bound group, a few additional differences emerge. Once again, we find that the families of the foreign academia-bound graduate students are noticeably more likely to regard advanced degrees highly than the families of the corresponding U.S. graduate students (40% vs. 20%). The U.S. group is rather more concerned with making a comfortable living (63% vs. 50%) and somewhat less concerned about helping the less fortunate at home and abroad (31% vs. 36%). By a fairly large margin, the U.S. group is more concerned about the extent of the involvement of the engineering profession with the military (28% U.S. group vs. 13% foreign group) and about the domination of the profession by white males (21% vs. 8%).

University-based recruiters might want to keep these differences in attitudes in mind. Although on university campuses engineers are not generally considered to be an especially liberal or idealistic group, among engineering graduate students it is the more liberal or idealistic ones who are more strongly attracted to academic careers. Recruiters could reinforce the sense that ac ademia would be a congenial environment for this segment of Ph.D. candidates, particularly the U.S. students.

Engineering Disciplines

Among all of our graduate student respondents, 22% had been, as undergraduates, chemical engineers, 11% had been civil engineers, 29% had been electrical engineers, and 24% had been mechanical engineers. (The rest had been in other engineering, physics, chemistry, mathematics, and yet "other.") The chemical engineering graduate students had the highest overall GPAs as undergraduates (3.43) and the civil engineers the lowest (3.21), with electrical engineers (3.37) and mechanical engineers (3.35) in between. In the undergraduate sample, the electrical engineering majors had the highest overall GPAs and the civil engineering majors the lowest. Given the pattern we have identified in which the ablest students tend to head towards doctoral degrees, research, and academia, it is important to examine the extent to which this pattern holds if disciplinary specialization is a variable. More than any other variable, we may suppose, disciplinary specialties have implications for types of post-educational activities and for employment opportunities.

It is interesting to note, then, that the graduate students in chemical engineering most frequently responded that they were in graduate school because they wanted to pursue careers in research (51% vs. 46% for electrical engineering students, and the rest were lower yet), though it was the electrical engineers who were much more likely to go to graduate school to learn more about their field (75% vs. 55% for chemical engineers). The chemical engineers were also least persuaded that the long-term economic rewards of graduate education would more than balance the costs (18% vs. 27% for electrical engineers, who were the most optimistic on this score). Chemical engineers were by far most likely to be pursuing the Ph.D. as their highest degree (as many as 80%, compared to 67% for electrical engineers, 66% for mechanical engineers, and 62% for civil engineers). It should be noted that even as undergraduates, those chemical engineering majors who planned to go to graduate school were considerably more likely than the other engineering majors to expect to obtain a Ph.D. (49% vs. 36% for mechanical engineering majors, 35% for electrical engineering majors, and only 19% for civil engineering majors). Even among the graduate students *not* pursuing doctorates, chemical engineers were most likely to hope to be doing research after they completed their education (82% vs. 74% for electrical engineers and only 55% for civil engineers) and to make significant contributions to knowledge (62% vs. 58% for electrical engineers, and only 44% for civil engineers) though with regard to teaching in general and university teaching in particular, chemical engineers responded similarly to all the other groups.

Our findings suggest that the chemical engineer respondents, who have the highest undergraduate GPAs of all the disciplines, are the most strongly disposed of the students in the various disciplines surveyed to obtain doctoral degrees and to go into research. One can speculate why this is so. It could mean that stronger students select themselves to chemical engineering because they are interested in research and believe that jobs entailing research are readily available; they may be attracted also by the nature of chemical engineering education, which has a more scientific orientation than other engineering disciplines because it incorporates chemistry. It could also mean that a temporary softness in the job market for those with bachelor's degrees convinces even high GPS students to go on for graduate degrees. By contrast, civil engineers, with the lowest GPA averages among the engineering disciplines, may find their specialty congenial because it does not require doctoral degrees and is not likely to involve research.

Type of Institution

Like the sample of engineering seniors, the graduate student sample was stratified by institutional or departmental quality and by institutional governance.

Effects of Quality/Research Intensiveness. We found no very significant effects on engineering seniors from quality differences in the departments in which they studied, and the same is true with regard to graduate students. As in the case of the undergraduates, and not surprisingly, the patterns of response of graduate students at the QRI-1 departments (the highest of our three quality groupings) resemble those of students with high GPAs and, therefore, also students who are headed for doctoral degrees. Graduate students in the QRI-1 departments are more likely to want to pursue careers in research and less likely to be concerned about their salary prospects. They are less likely to have had no attractive full-time job offers (19% for QRI-1 vs. 26% for QRI-2 and 34% for QRI-3) and more likely to have been offered good financial aid packages (75% for QRI-1, 73% for QRI-2, and 67% for QRI-3). In dollar terms, the average monthly stipend received by students in QRI-1 departments, excluding tuition, was $799; in QRI-2 departments, $707; and in QRI-3 departments, $595. The difference here between the highest and the lowest is quite considerable.

Not surprisingly, QRI-1 students are much more likely to be pursuing doctorates than QRI-3 students (53% vs. 26%), but, and this may be somewhat more surprising, there is not very much difference in this regard between

QRI-1 and QRI-2 graduate students (53% vs. 46%). Along the same lines, QRI-1 graduate students are the ones most likely to be headed for careers in research and teaching.

We can only conclude here that the best students go to the highest quality departments and, therefore, it is impossible to disentangle the influence of their academic ability from the influence of the departments without careful analysis of the performance of high-ability students in QRI-2 and QRI-3 departments or weaker students in QRI-1 departments. The two factors, academic ability and departmental quality, may reinforce each other, but we lack the evidence to demonstrate this.

Effects of Governance. The same sort of patterns, but much less pronounced, can be found if we compare the responses of students in private and public institutions. Those in private ones had slightly higher GPAs (undergraduate GPAs of those in private institutions were 3.39 vs. 3.34 in public ones). The graduate students in the public universities were more likely to take into account, among the economic factors in their decision to attend graduate school, that the living expenses would be fairly low (64% of those at public institutions cited this as very or somewhat important, compared to 37% at private institutions). Students in private universities more frequently pursued doctoral degrees as their highest degrees (74% vs. 66%), and those in public universities more frequently felt that a doctorate would not help them to do the kind of work they want to do (55% vs. 48% in private institutions) and that a doctorate was too theoretical and research-oriented for them (54% vs. 43%). Finally, those in private institutions were slightly more inclined to go into research (74% vs. 71%) and into university-level teaching (49% vs. 44%). Once again, the graduate students in the private institutions are of higher quality, but whether it is their superior ability as measured by undergraduate GPAs, or the influence of the institutions that affects their academic and career decisions cannot be established on the basis of our data.

Conclusions

The factors that come into play in the decisions of engineering seniors and graduate students to continue or not to continue to graduate study are, not surprisingly, quite complex. From the time they are undergraduates, students seem to relate their academic goals and their career aspirations quite directly to their academic performance. Among undergraduate students, those who perform well academically and develop a sense of academic competence are more likely to continue to graduate study. The choice of engineering major is also a factor, with students in more research-oriented disciplines (chemical and electrical engineering) showing higher interest in graduate work. Among graduate students, the students with GPAs in the highest quartile tend to pursue doctoral degrees and to prefer research and university teaching, as compared to students in lower quartiles. (Inescapably, in dealing with student quality, we have resorted to the use of self-reported GPAs as the basis of our analysis.) However, there are clearly many undergraduates, 90% of whom are U.S. citizens, who are capable of pursuing graduate study and do not, in fact, choose to do so. Only 19% of our senior student respondents were planning to go to graduate school in engineering immediately after graduation.

Economic factors, while important in student decisions, are far from the whole story. Many students, at both academic levels, give evidence of commitment to making significant practical or intellectual contributions through their work in engineering. That does not mean that a quite high proportion of them do not aspire to making a comfortable living. We may assume that their academic and career decisions are being made in the context of their assessment of fluctuating job opportunities (which we could not factor into our study) and that the better students are in a better position to follow their preferences — for engineering specialties, for types of activities, for careers — without too much concern for changes in the engineering labor market.

It is worth noting that among these better students, research has great appeal, but the appeal of university teaching is appreciably less strong.

Whatever the job market and whatever the graduate students' career aspirations, foreign students are clearly more likely than U.S. students to feel that they have a much better chance of achieving their goals if they obtain doctoral degrees. Since 83% of our foreign graduate student respondents did their undergraduate work before they came to the United States, it is not easy to make direct comparisons of the undergraduate academic performance of the foreign and U.S. groups; the evidence based on GRE scores is ambiguous because foreign students score so much lower on the verbal component of the test. In any event, there is no reason to believe that, on average, the academic quality of foreign students is higher than that of U.S. students. At the present time, however, only 60% of our U.S. graduate student respondents as against 80% of foreign graduate students found it necessary or attractive to obtain a doctoral degree. We may conclude that a considerably higher proportion of U.S. students are quite capable of doing doctoral work and that they might well do so if it were necessary to assure desirable occupational outcomes for them. As long as non-academic employers convey to U.S. engineering students that they need not worry about jobs as long as they have B.S. or M.S. degrees, a substantial proportion of them will choose to avoid the cost and travail of obtaining a doctoral degree.

Our study reveals several points at which new approaches may be effective in increasing the flow of U.S. students into graduate programs in engineering and, more specifically, into doctoral programs. Some of these should be considered by engineering educators, some by employers, and some by policymakers and the concerned public.

(1) A substantial proportion of seniors are dissatisfied with their undergraduate programs. The major complaint is that programs are too demanding and that students feel burned out. It seems desirable to reexamine the undergraduate curriculum to find ways in which this situation might be ameliorated. Given the large number of engineering undergraduates, the small proportion that go on to graduate study, and the relatively high percentage of our graduate respondents seeking doctorates, a relatively small increase in the flow into graduate school could produce a substantial increase in doctorate production.

(2) We have some evidence that those undergraduates who become involved in pre-professional activities (tutoring, student membership in chapters of science or engineering associations) tried to become involved eventually in advanced study. Our data also indicate that working on campus in research projects or involvement in coop programs is positively correlat-

ed with going to graduate school. Efforts to encourage such activities may well be helpful in recruiting more students, including women and minorities, into the academic pipeline.

(3) Attention might well be given both by engineering educators and employers to the students who say they do not wish to go to graduate school immediately but might do so later. Presumably, this means that there are numbers of engineers in industry who might be enticed into Ph.D. programs if given proper incentives and encouragement by their employers and by the engineering schools. Bringing this about will require cooperative efforts between industry and universities so that industry will not feel it is being deprived of needed personnel.

(4) The greater amount of encouragement with regard to doctoral work in engineering that foreign students get from their families suggests that there may be a need in the United States to place a higher value on both educational achievement and the engineering profession. Larger stipends for graduate study may, in this regard, have both practical and symbolic significance.

(5) Our results indicate that recruiting efforts by graduate programs are not viewed as being very visible or effective. More and better efforts by engineering schools are needed. Since our data show that positive decisions about graduate education are related to perceptions that advanced degrees lead to stimulating, challenging work, it would be desirable to emphasize this prospect in recruiting materials.

(6) If faculty gave undergraduates more encouragement to pursue graduate education and related careers, more students might do so. In this regard, broadening the engineering faculty base to include women and minorities could be a key step in increasing the flow of students from these groups into graduate programs. Such a step could also serve to counter the impression that is held disproportionately by these groups that engineering is a profession for white males.

(7) Although U.S. female engineering students appear to be somewhat more people-oriented than their male counterparts, they have many of the same aspirations, impressions of their engineering program experiences, and decision priorities as do males. Our survey also indicates that a higher percentage of daughters of engineers tend to enter the engineering profession than do the sons of engineers, a result that might prove useful in identifying future female engineers. A recent survey by Baum,[7] exclusively directed towards professional women engineers, provides valuable additional information.

(8) Because the small number of responses we received (in spite of oversampling) from U.S. black, Hispanic, and Native American students limited the value of analyzing results for each of these groups separately, we aggregated these responses into one underrepresented group and compared them with the responses of two other groups: Asian-Americans and U.S. whites. Even this aggregation leaves much to be desired statistically; any conclusions are, at best, informed guesses and, at worst, hypotheses worthy of further study — study that is badly needed. Four points that emerge are that members of underrepresented groups are: a) more limited by financial need in going to graduate school; b) get less encouragement to go from various parties; c) have fewer engineering relatives; and, d) are less oriented towards research than members of the other groups. Yet the desire and motivation of underrepresented respondents to go on for advanced engineering study are equal to or greater than those of their Asian-American and white counterparts.

There is reason to believe that engineering students, generally, tend to pursue a doctorate if they surmise that they might not be able to obtain desirable jobs without it, as seems to be the case with foreign graduate students. The fact that young U.S. engineers can get well-paying jobs with bachelors and master's degrees looms as a major obstacle to expanding the supply of U.S. citizens with doctorates in engineering. This situation raises a number of fundamental questions that require attention. How are engineers at each level — B.S., M.S., and Ph.D. — utilized? Do we really need more engineering Ph.D.s who have gone through a research-oriented experience modeled after that in the natural sciences? What are the relative contributions of engineers at each degree level to U.S. productivity and competitiveness? How rewarding do graduates at each of these degree levels find their assignments and careers?

Depending upon the answers to these questions, a variety of policies are possible. Two of these are: 1) to require a a master's or other advanced engineering degree as the first professional degree; and, 2) to deemphasize somewhat the research-oriented engineering doctorate, creating a more clinically-oriented or practically-oriented graduate educational experience in parallel with the traditional Ph.D. Even if the doctoral degree were deemphasized, we would expect that foreign students will continue to pursue the Ph.D. degree disproportionately, in order to get whatever advantage they can from it in the job market. The engineering profession has been struggling with these important questions for decades. The time may be ripe to give them a new and hard look.

References

1. National Research Council, "Foreign and Foreign-Born Engineers in the United States," National Academy Press, 1988, p. 2.

2. Barber, E.G. and R.P. Morgan, *Science*, Vol. 236, 3 April 1987, pp.33-37.

3. Barber, E.G. and R.P. Morgan, *Boon or Bane*, New York: Institute of International Education, 1988.

4. For methodological details, see W.P. Darby, R.P. Morgan, L.J. Sallmen-Smith, J.A. Smith, and M.E.A. Stephens, "Factors Affecting the Supply and Composition of the Graduate Student Population in U.S. Engineering Schools: Sample Selection and Survey Procedure," Report no. CDT-88-2, Center for Development Technology, St. Louis: Washington University, November, 1988.

5. Jones, L.V., G. Lindzey, and P.E. Coggeshall, Eds., "An Assessment of Research-Doctorate Programs in the United States: Engineering," National Academy Press, 1982.

6. *Engineering Education*, Vol. 77, No. 6, March 1987, p. 329 ff.

7. Baum, E., "What's It Like to Be a Woman Engineer?" *Engineering Education*, Vol. 16, No. 5, December 1989, p. 1.

List of Tables

List of Figures